microbe (mī-krōb′)

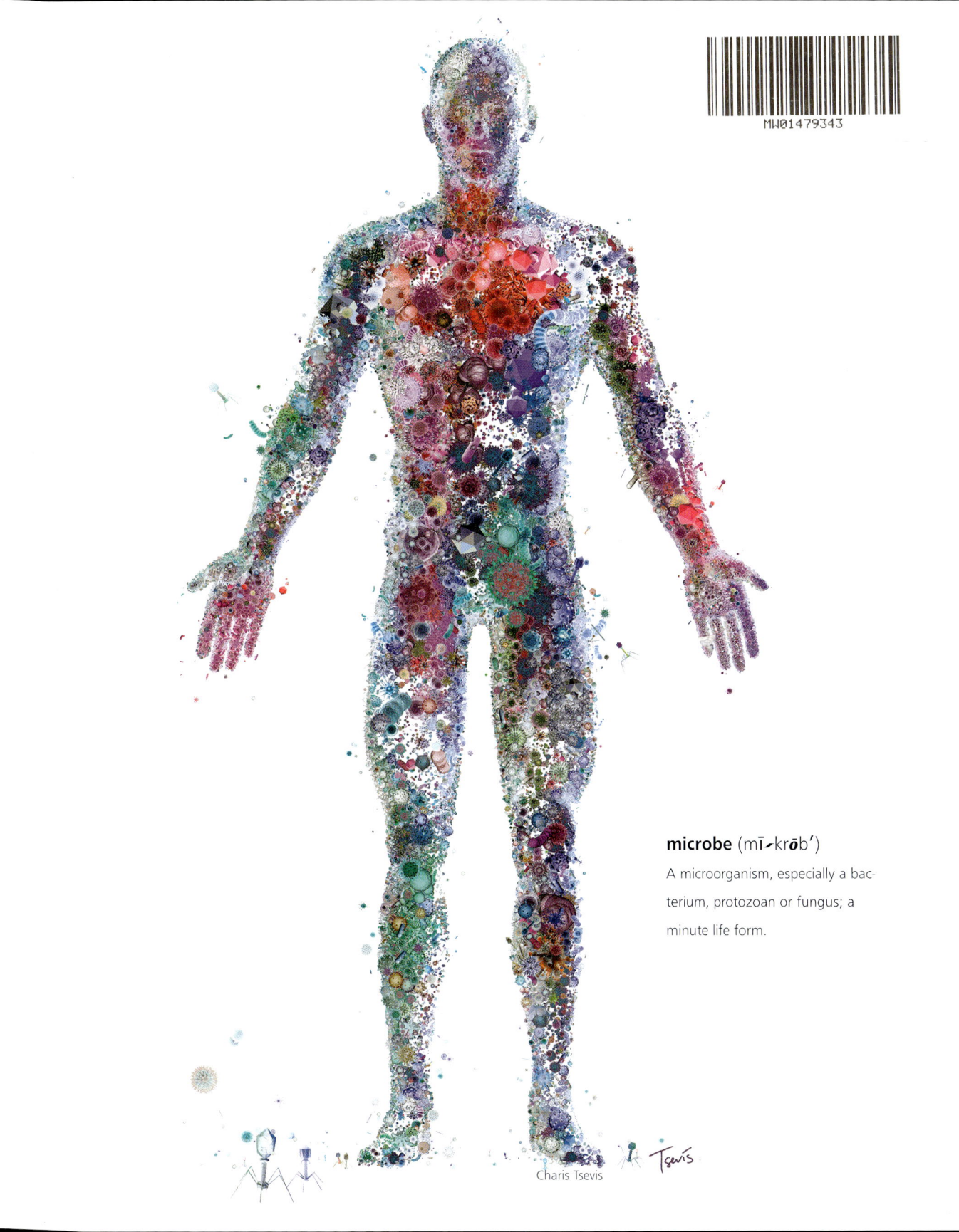

A microorganism, especially a bacterium, protozoan or fungus; a minute life form.

Charis Tsevis

Science Source

Mark Thiessen/National Geographic

Science Source

Ice caves, sulfur pools, alpine forests and barrens of treeless prairie make Idaho a revealing state of biotic extremes. New discoveries in microbiotics are helping Idahoans preserve food, conserve soil, purify water, graze cattle and explore the human connection to single-celled organisms.

Pictured, from left: West Nile, a virus vulnerable to mosquito-killing fungi (chapter 8); searching for ancient bacteria under an ice sheet (chapter 3); *Giardia* colonizing a human intestine (chapter 5); Louis Pasteur, father of microbiology (chapter 1).

Seung-Hwan Oh

Fungus devours the photographer's film in Seung-Hwan Oh's decomposition of a portrait distorted by microbes

Idaho Microbes

how tiny single-celled organisms can harm, or save, our world

by
Steve Stuebner
with Todd Shallat

Boise State University
School of Public Service and Division of Research and Economic Development

B

contributors

Todd Shallat
editor

Colleen Brennan
associate editor

Adele Thomsen
graphic designer

Brian Marinelli
education editor

Greg Hampikian
science advisor

Corey Cook, Dean
School of Public Service
Boise State University

Mark Rudin, Vice President
Research and Economic Development
Boise State University

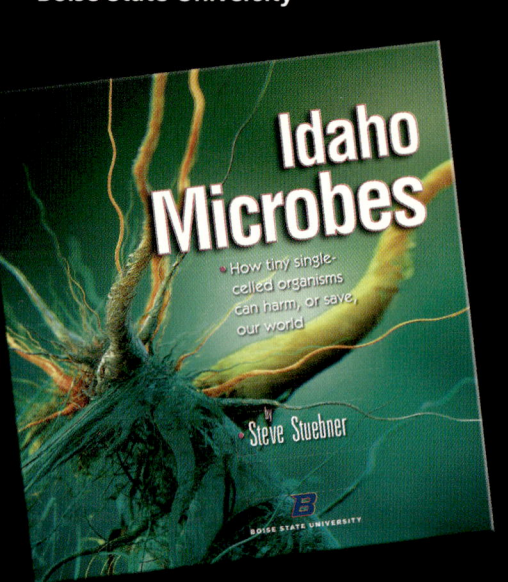

About the cover: Microscopic filaments tangle and subdivide in Václav Pajkrt's "Growth of Cubic Bacteria." Right: A phage virus rockets from bacterial host to host in Russell Kightley's "Phage Dreaming 1."

Russell Kightley

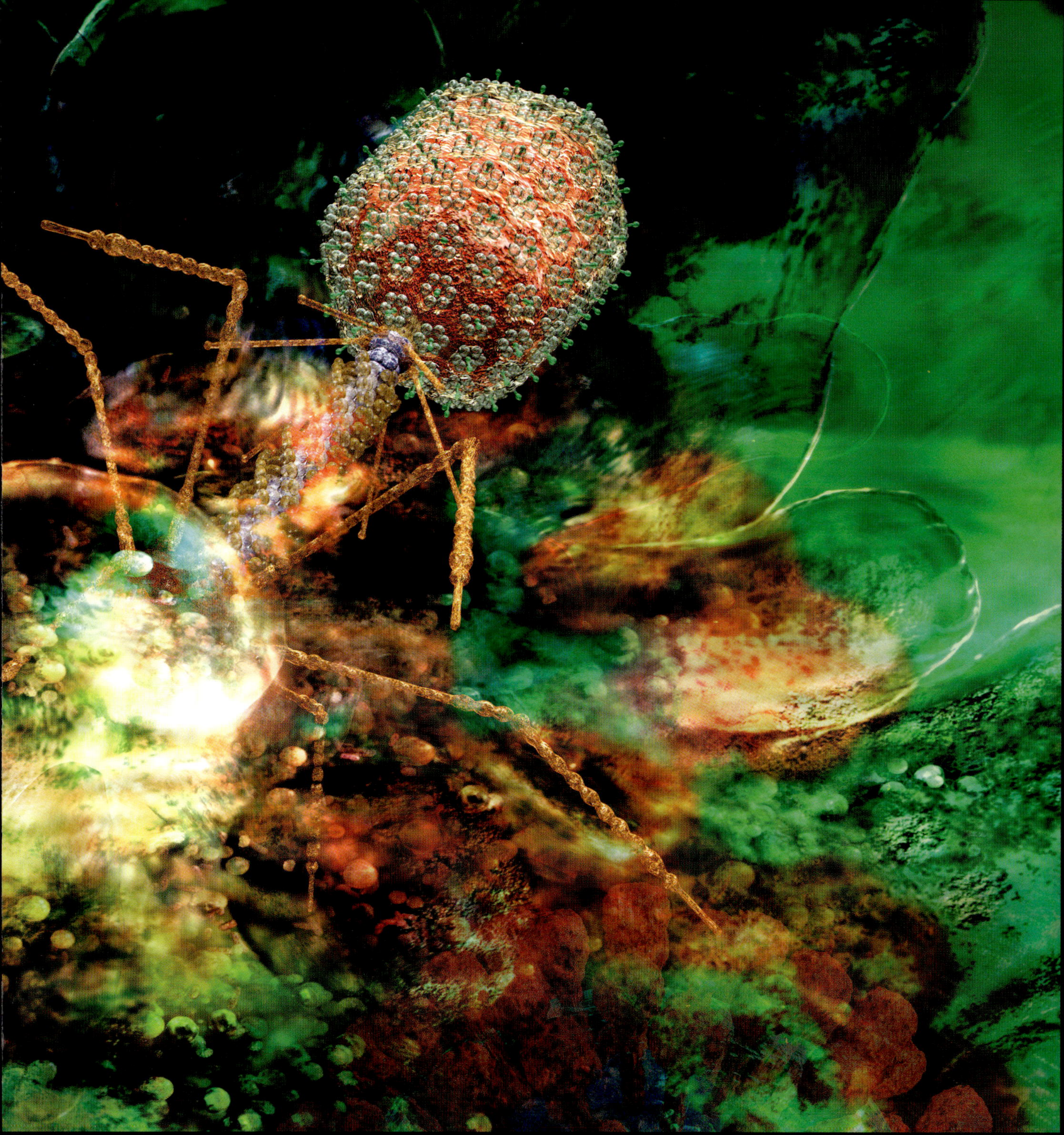

Centers for Disease Control and Prevention

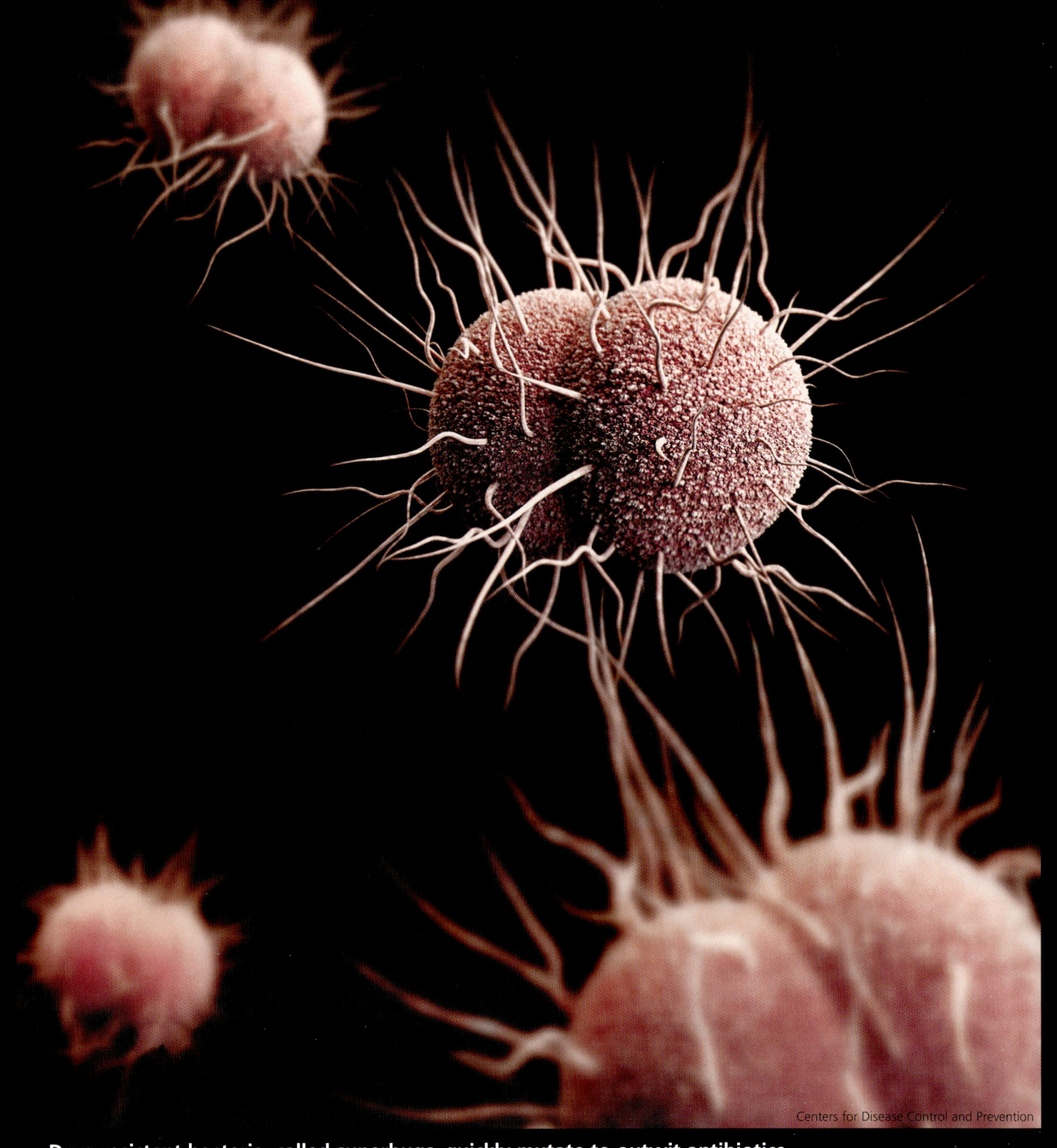

Drug-resistant bacteria, called superbugs, quickly mutate to outwit antibiotics.

Contents

Foreword 13

1. Modern-Day Microbe Hunter 16
2. Wondrous Mycelia 38
3. Methane from Microbes 68
4. The Magic of Yeast 90
5. Giardia Lurks 112
6. Forest Epidemic 132
7. Guzzling Crude 160
8. Home Inside a House Fly 184
9. Freddy Fungus Meets Alice Algae 208
10. Saving the Soil 232

Selected Sources 258

Index 262

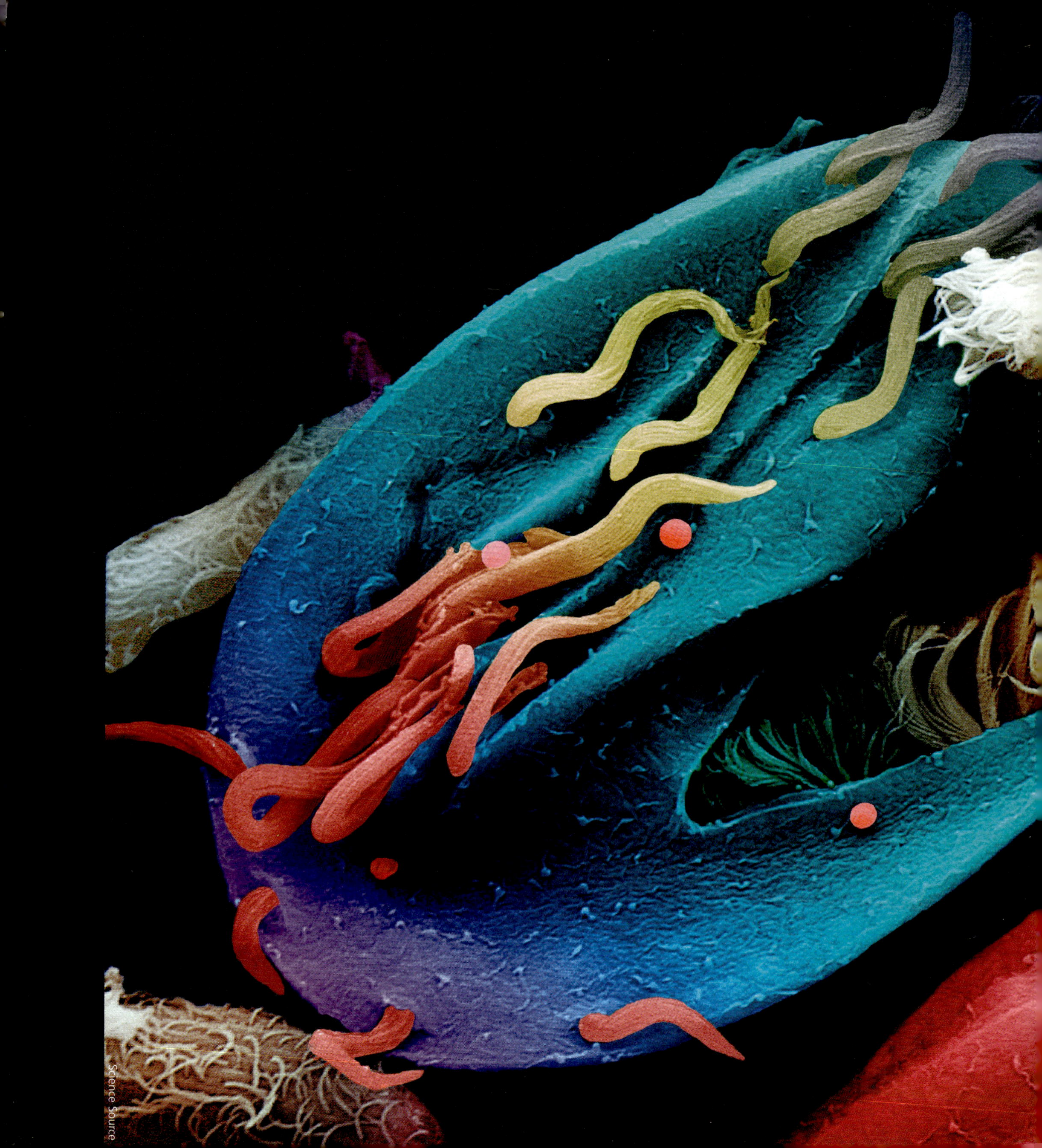

Science Source

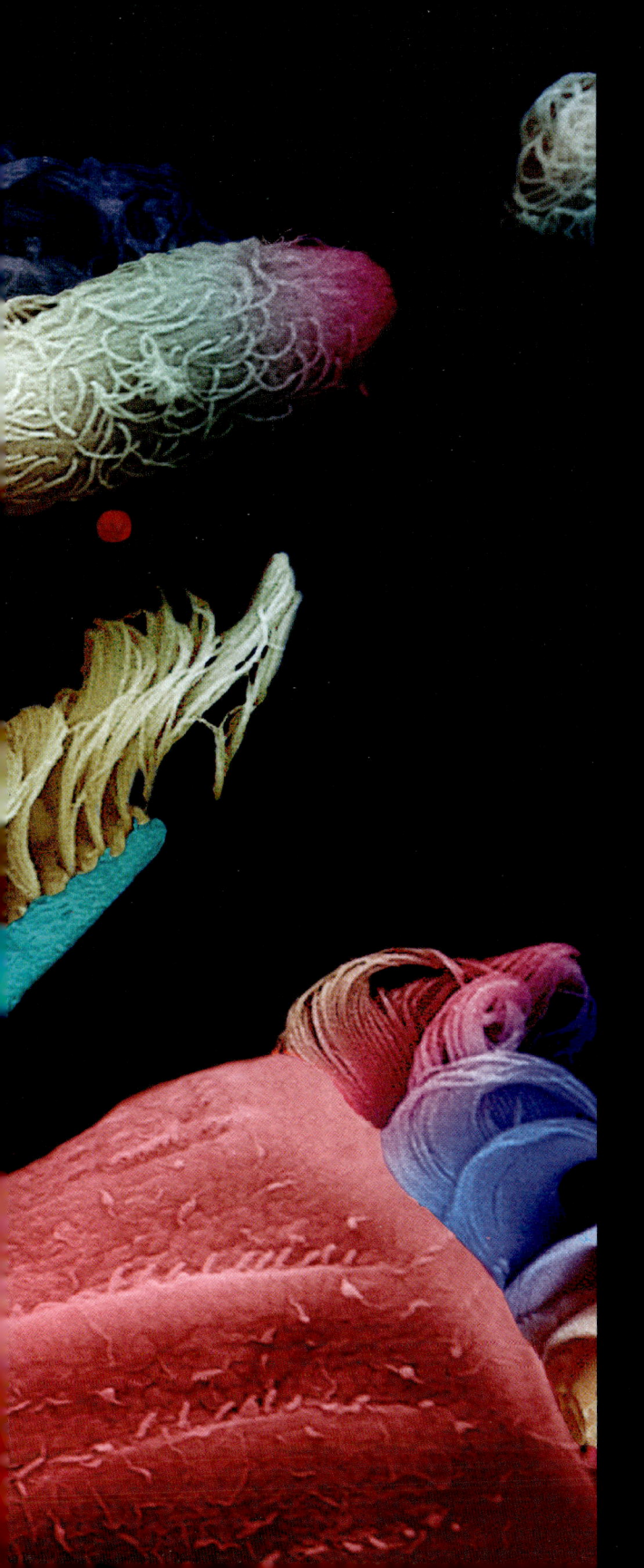

Boise State University
School of Public Service and Division of Research and Economic Development

Ordering:
http://sps.boisestate.edu/publications/
spspublications@boisestate.edu
(208) 426-1350

Editorial:
tshalla@boisestate.edu
(208) 761-0485

Mailing:
1910 University Dr.
Boise, ID 83725-1925

All rights reserved.
Hardcover edition
ISBN 978-0-9907363-0-1
2015

Euplotes protozoa scavenge ponds for bacteria. Propelled by slender hairlike flippers called cilia, the microbe can swim, walk or ride by clinging to other species.

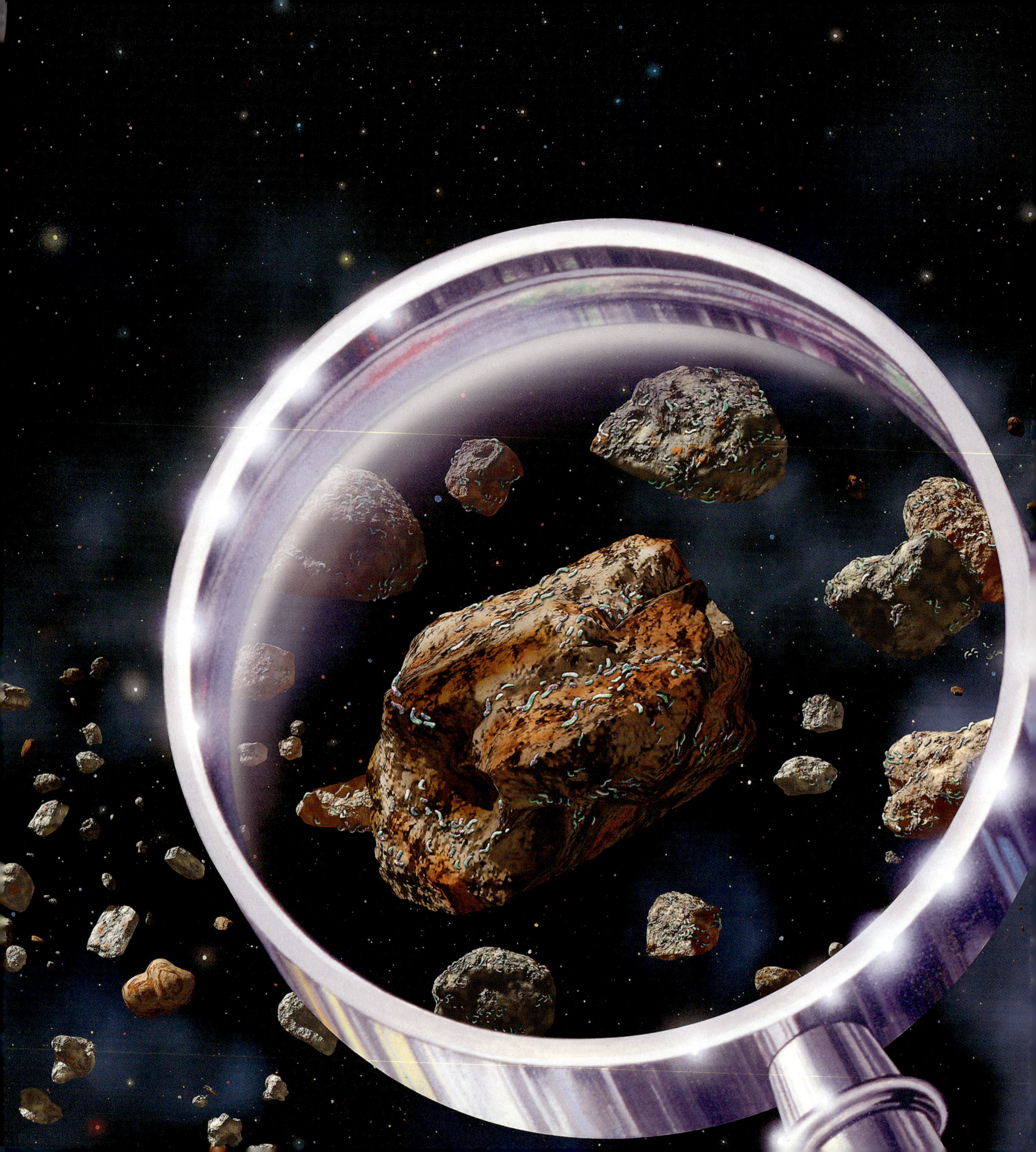

Foreword

If microbes wrote biology books, human beings would be a small footnote, a recent development—not very numerous, but generally useful as incubators. Our constant temperature, blood glucose, oxygen and CO_2 are especially rich breeding grounds for all kinds of small communities, and the hollow tube running from our mouth to our colon is a safe and fertile ground for microbes of all sorts. From the moment that we enter the birth canal to our eventual decomposition, our microbial inhabitants outnumber our own cells and perform essential services, including nutrition and defense. Evolutionarily, all life came from microbes, and in death, to microbes we all return.

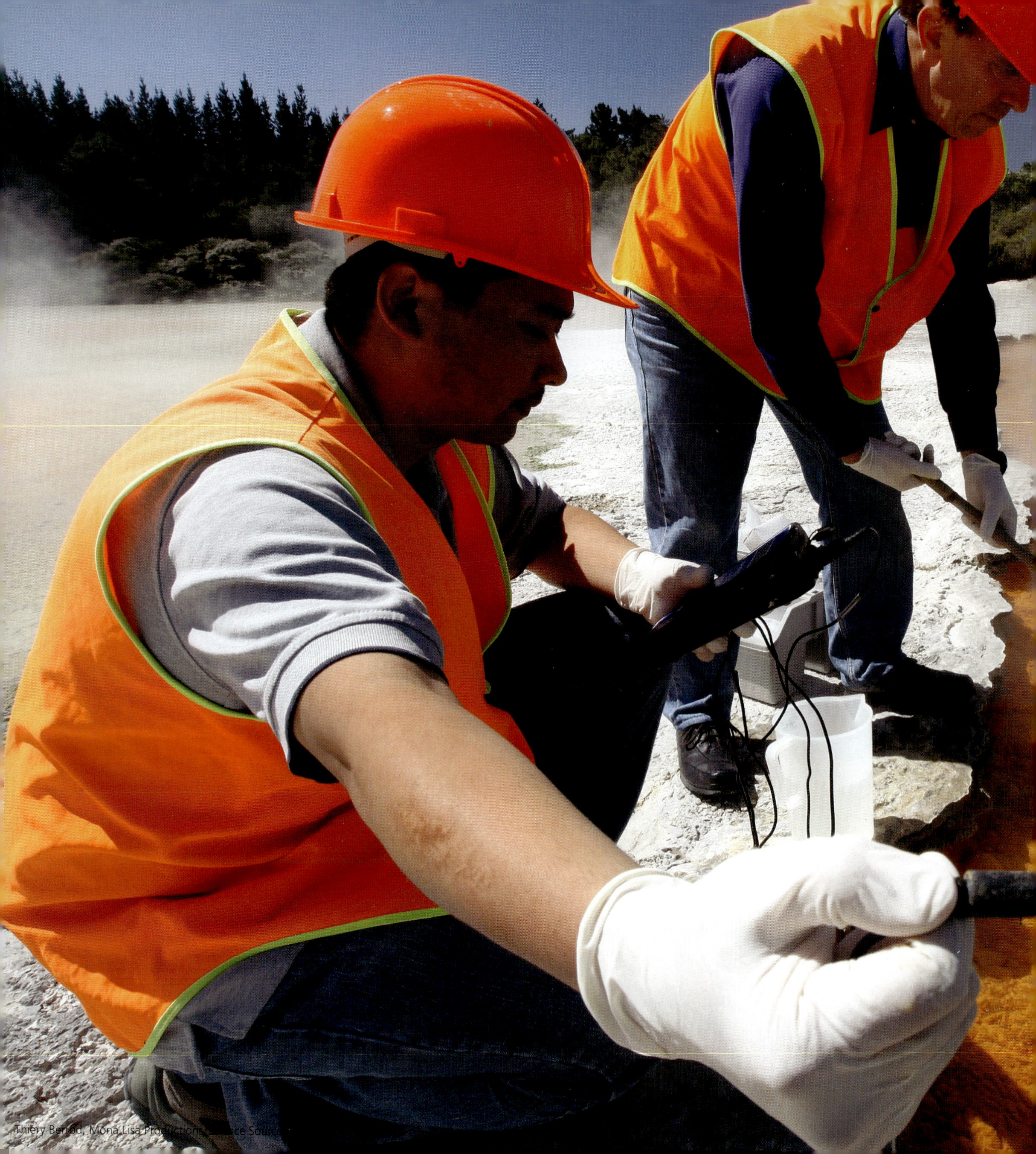

Thiery Berrod, Mona Lisa Productions/Science Source

The good news is that as long as we are faithful to the partnership, the microbes will allow us to flourish. The corollary, however, is that if we alter conditions in the environment or within our own bodies too much, these helpful companions will overtake and destroy us. They were here first, and they will certainly be here last. Remember that while human beings have been around for just a few million years, microbes have a history that dates back at least 3.5 billion—and that is just on Earth. Since our planet is roughly 4.5 billion years old, the partnership between bugs and Earth goes back to nearly the beginning—and it worked quite well without us for a very long time.

As we now begin to look at our universe with not just telescopes but also microscopes, we may find even older life on asteroids or other planets, and these long-successful organisms could date back further toward the beginning of our universe, which is approximately 15 billion years old. Some scientists now believe that planets like Earth are seeded by interplanetary microbes (panspermia) and that these small, living diversity-engines are capable of eventually generating all the complex communities found on a planet like ours.

Diversity and complexity are two themes that frame our eco-adventure through Idaho's microbial habitats. We invite you to travel with Boise nature writer Steve Stuebner as he finds microbes working both for and against civilization. You'll visit Idaho's western steppe in the Northern Rockies and learn how microbes labor in farming and industrial fermentation. You'll see how microbes bind soil in sagebrush rangelands, convert cow manure to valuable methane, plague forests with invasive species and cleanse groundwater polluted by oil. Seeing the unseen, we hope that you'll never look the same way at a clod of dirt or a glass of water. Perhaps you'll be inspired to join the hunt and discover microbial species in your own environment. Just remember, they discovered you first.

Greg Hampikian, Ph.D.
Department of Biological Sciences
Boise State University

Orange-colored mats of heat-seeking microbes thrive in Idaho-Wyoming geothermal hot springs. Previous: Scientists speculate that life on Earth may have originated from microbes hitching rides on space debris.

Russell Kightley

modern-day microbe hunter

Dr. William Bourland
citizen scientist

1

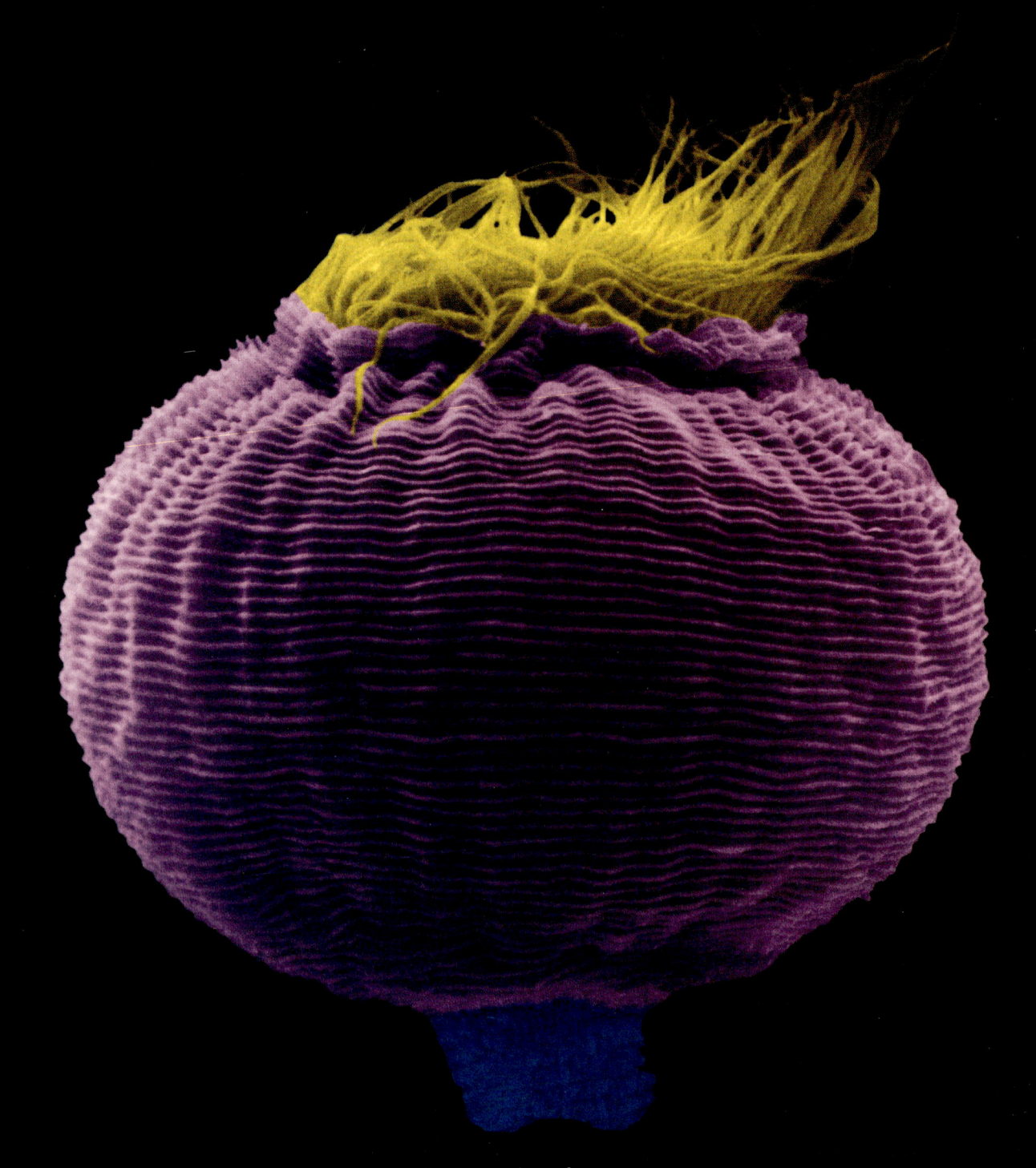

Protozoan Vorticella

Left: Tiny hairs called cilia propel thousands of species of ciliated protozoa through Idaho puddles and ponds. *Cilium* is Latin for eyelash.

Below: Miniature marvels of translucent nature beckon William "Bill" Bourland, M.D. Previous: Artist Russell Kightley mutates microbial patterns in "Microscope Dreaming 4."

William "Bill" Bourland is fascinated by ciliated protozoa—very sophisticated single-cell organisms that live in the water and the soil. Retired early after a 22-year career as a vascular and thoracic surgeon, Bourland spends his free time hunting for new species of ciliated protozoa in various mud puddles in the city of Boise. It's not your typical hobby, certainly. But Bourland has had a lifelong interest in tiny organisms that can only be seen and understood through the lens of a high-powered microscope. For him, it's a passion, a chance to discover new species that haven't been found anywhere in the world, species that could help us understand more about how life evolves, spreads and organizes into complex communities.

I go microbe hunting with Bourland on a warm, hazy day in August. Dressed in a light green T-shirt, cargo shorts and sandals—typical Boise summer wear—Bourland drives a green Subaru to his "hunting" area. We arrive at the deep green finely manicured lawns surrounding the Veterans Administration Medical Center in the Fort Boise area, where Bourland used to perform surgery. He parks next to a section of lawn that's interspersed with young maple trees, bordered by concrete curbs. It's a spot that most people would walk by and see nothing at all. But in his mind, Bourland can see microscopic creatures living underneath the bluegrass.

He bounces out of the car, lifts the tailgate and produces his weapon—a turkey baster with a bright red bulb on the top

John Kelly/Boise State University

modern-day microbe hunter 19

of it, the same instrument your grandmother might have used to prepare the turkey for Thanksgiving dinner. He ambles up near a sprinkler head and pushes down with his sandal, and I hear the water-soaked sod go "squish."

"Oh yeah, I think we can get some samples out of here," he says. "It doesn't have to be any kind of exotic special place. Once you start looking around, you start seeing these little mud puddles everywhere. It's all there waiting to be found."

Bourland sticks the turkey baster deep into the grass, squeezes the bulb and the clear plastic cylinder fills with a brownish-clear liquid. "Got 'em!" He grabs a glass jar from the back compartment of the Subaru and squirts the liquid into the jar. He pushes his foot down again, inserts the baster and draws more liquid until the jar is three-quarters full. We can already see larger creatures swimming around inside. "That'll be plenty," he says.

This all reminds me of science class in elementary school, when we sampled water from a pond behind our school and put samples on slides to see what kind of single-celled organisms lived in the pond, species that we couldn't see with the naked eye. That's what we're going to do at Bourland's house. We drive back to his home in the foothills of Boise's hip North End and go into the basement, the location of his home laboratory. In the

Left:
German toy makers helped popularized brass microscopes with adjustable lenses and mirrors. Botanical slides mounted on cardboard became, by the 1870s, standard accessories in medical schools.

Below:
Science quickly evolves and so does biological classification. Modern-day microbe hunters mostly classify life in three domains with six kingdoms. Ciliates, being species of protozoa, inhabit the kingdom of Protista.

center of a black countertop sits one of Bourland's prized possessions, a research-quality Zeiss microscope. The microscope has a digital camera mounted on top that's connected to a Sony computer monitor next to it. He places a drop of water from the glass jar on a clean specimen slide, drops a glass cover over the bead of water and loads it into the Zeiss microscope.

Bright halogen lights illuminate the slide. Bourland removes his wire-rimmed glasses and peers into the scope, while his hands move quickly, adjusting the power and focusing in on the critters that come into view. I can see Bourland's eyes grow wide and dart around as he looks at the water sample. We see tiny round worms (nematodes) swimming around, and he zooms in closer, and we see the even smaller rotifers, which are multicellular organisms (metazoans). He zooms in some more, and we start to see ciliated protozoa swimming around. These are Bourland's target species, each a complete animal in just a single cell. "Some ciliates are anchored to soil particles or bits of plants," he says, "but I like to look at free-swimming mobile forms."

We see a kidney-shaped ciliate called *Colpoda*, a relatively common terrestrial genus. Bourland zooms into the creature. Our exploration of the cell's anatomy begins on the blue computer screen as he points out the organism's nucleus, then its tiny hairlike organelles that it uses for mobility, its opening in the outer membrane for ingesting food and another opening for discharging waste and more.

All ciliate species possess tiny hairlike organelles known as *cilia*—hence, the name of this group of protozoa. The cilia look similar to flagella in other

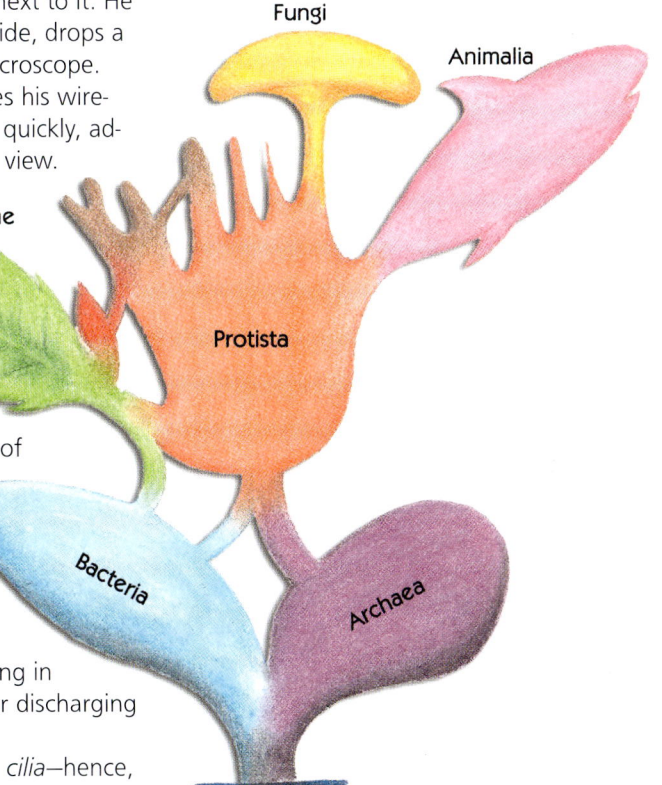

Maulaucioni y Doridi

Steve Gschmeissner Fine Arts America

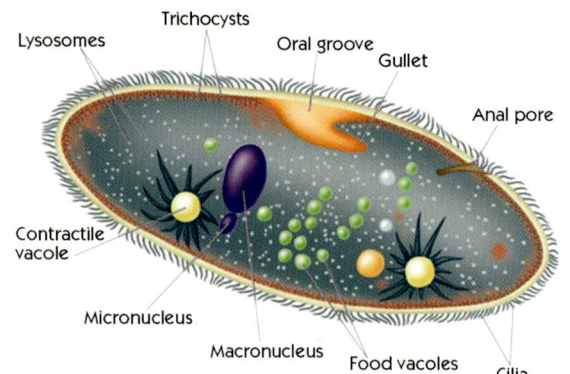

Lysosomes
Trichocysts
Oral groove
Gullet
Anal pore
Contractile vacole
Micronucleus
Macronucleus
Food vacoles
Cilia

eukaryotic species, but the organelles are shorter and much more numerous, covering the outer membrane of the cell. Ciliated protozoa use the cilia for locomotion, whether it be swimming, crawling, attachment, feeding or sensation.

"This is how I go about looking at things," Bourland says. "The simplest is the first step: get the specimen and see what's in it. Then I begin studying it in more detail with different optical techniques and classify it. My interest is to find new organisms that have never been described before."

So far, Bourland has discovered three unique terrestrial ciliates, *Bryophrya gemmea*, *Etoschophrya inornata*, and *Agolohymena aspidocauda*. He discovered a fourth, *Bryophryoides ocellatus*, just a few weeks ago, in a puddle on the well-irrigated green lawn behind the Boise Little Theater. He likes to draw samples from mud puddles next to Cartwright Road, various irrigation canals,

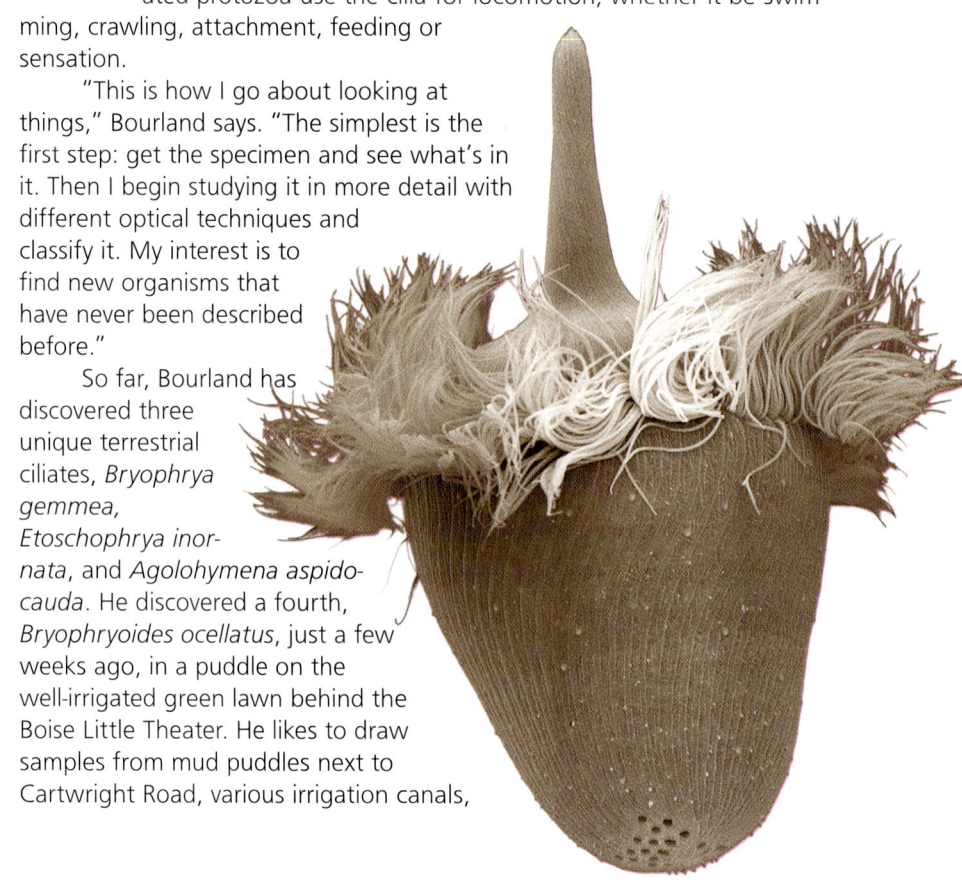

22

Left:
The pond-dwelling *Monodinium balbianii* feeds on bacteria. Buoyant and predatory, it flaps its finlike ciliaryvgirdle, propelling the microbe toward food.

Below:
NASA's Felisa Wolfe-Simon, an astrobiologist, hunts microbes in a core sediment sample from Mono Lake in the High Sierras. Lake organisms that metabolize arsenic may resemble extraterrestrial life.

several parks in Boise and farm ponds. He's also collected samples from the desert flats where raptors like to catch mice in the Morley Nelson Snake River Birds of Prey National Conservation Area, south of Boise. When traveling the world with his wife, Debra, he's collected samples in Finland, Brazil and Borneo.

On a website called "Micro*scope," maintained by Arizona State University, Bourland has shared more than 1,700 digital images of about 650 species of ciliates since 2004. He's met the world's top ciliatologists and collaborates with them on a regular basis. Although he went through a lengthy amount of schooling and training to be a surgeon, he's had to learn a whole new field of scientific discovery, a field called *protistology*, the study of *protists*—single-

modern-day microbe hunter 23

celled organisms, including unicellular algae and protozoa. He's had to learn a whole new vocabulary, so he can speak and write the language of protistologists when publishing his findings in scientific journals.

Bourland is a "citizen scientist," a category of intriguing people who have made huge scientific discoveries dating back to the 16th and 17th centuries. The first "microbe hunter" was Antonie van Leeuwenhoek, a Dutch carpenter who knew how to cut glass and invented the best microscope available in the mid-1600s. With the microscope, Leeuwenhoek began documenting single-celled organisms. He identified various forms of bacteria, free-living and parasitic microscopic protists, sperm cells, blood cells, microscopic nematodes and more.

In 1857, while working with yeast in Paris, microbiologist Louis Pasteur made the connection between airborne organisms and human disease. In doing so, he had to identify, classify and characterize many bacterial species in detail. He learned that some bacteria are good, even helpful, to human beings. And some bacteria are bad, causing infections such as tonsillitis, pneumonia and diarrhea. Pasteur came up with the first vaccines for rabies and anthrax, and he's well known for inventing *pasteurization*, a method for slowing the rate of spoilage for milk and wine.

Bourland is inspired by these early microbiologists, among many others, knowing how their discoveries have had such a profound impact on humans. One of his heroes is Alfred Kahl, a German high school teacher who began studying ciliates in mid-life, in the early 1900s. Kahl discovered 17 new ciliate families, 57

Vanity Fair

Left:
Louis Pasteur, microbiologist, transmitted rabies from rabbit to rabbit in experiments that pioneered vaccines. Pictured: lithograph of Pasteur from London's *Vanity Fair*, 1887.

Right:
Dutch tradesman Antonie Philips van Leeuwenhoek (1632-1723), with brass microscopes of his own invention, was the first to scientifically document single-celled organisms.

genera and about 700 previously unknown species. It was a time ripe for discovery.

In this book focusing on a diverse set of microorganisms, a number of leading experts, including Bourland, Boise State University professors Merlin White and Kevin Feris, and many others, will teach us about these tiny creatures, explain what makes them tick and how they impact society. Chris Florence of Sweet Valley Family Farms will show us how to find wild edible mushrooms growing in the cool, moist forests of Idaho, natural food that can turn an average meal into an unforgettable feast. On a broader level, he'll explain how a vast network of mycelia underpin the forest and provide a life-support system for many plant species.

Feris will culture microbes that feast on petroleum products and help us

modern-day microbe hunter 25

Below:
Edward G. Robinson starred in a 1940 Warner Brothers biopic about the German microbiologist Paul Ehrlich. Nazi Germany banned the film because Ehrlich was a Jew.

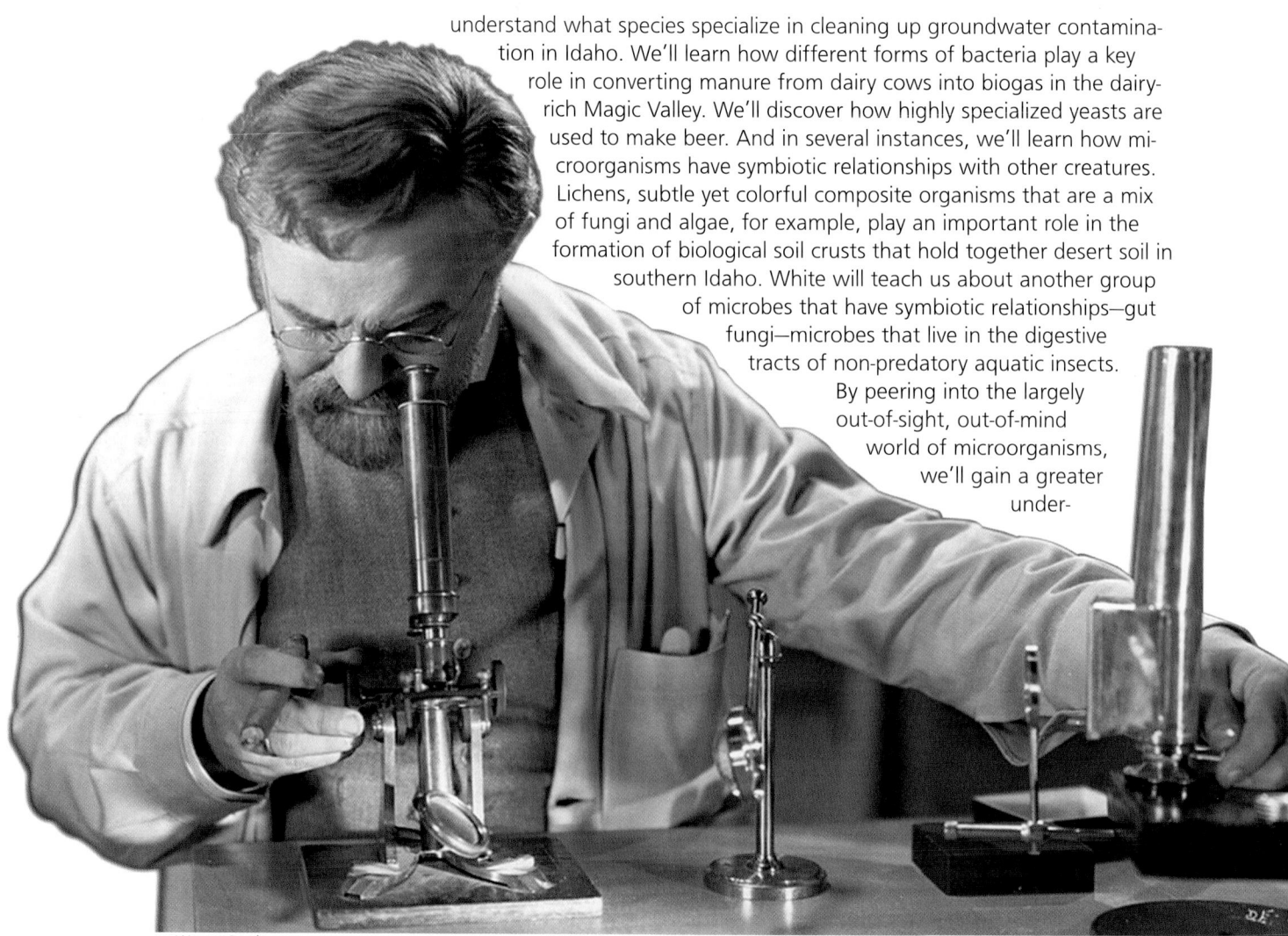

understand what species specialize in cleaning up groundwater contamination in Idaho. We'll learn how different forms of bacteria play a key role in converting manure from dairy cows into biogas in the dairy-rich Magic Valley. We'll discover how highly specialized yeasts are used to make beer. And in several instances, we'll learn how microorganisms have symbiotic relationships with other creatures. Lichens, subtle yet colorful composite organisms that are a mix of fungi and algae, for example, play an important role in the formation of biological soil crusts that hold together desert soil in southern Idaho. White will teach us about another group of microbes that have symbiotic relationships—gut fungi—microbes that live in the digestive tracts of non-predatory aquatic insects. By peering into the largely out-of-sight, out-of-mind world of microorganisms, we'll gain a greater under-

Warner Brothers

Left:
Ehrlich's legacy in microbiotics includes cancer research and a diphtheria serum. Pictured: Nobel Prize commemorative medal and a 200 Deutsche Mark banknote.

Below:
Microbe hunter Paul Ehrlich (1854-1915) pioneered the "magic bullet" medications that target disease-causing organisms. His Frankfurt laboratory developed the first effective cure for syphilis, a precursor of chemotherapy.

standing for how these creatures impact society in Idaho and beyond—in positive and negative ways. The people who bring these stories to life will hopefully inspire others—youngsters, college students or even adults whose careers have gone stale—to delve deeper into the field of scientific discovery, just as Bourland has done, to help unravel the mysteries of life on earth. In Bourland's mission to discover new species of ciliated protozoa, he found a ciliate, *Puytoraciella dibryophryis*, in a mud puddle in Ann Morrison Park, near the Boise River. This was a ciliate that he hadn't seen before, so he began poring through the literature to determine its uniqueness. It turned out that *P. dibryophryis* had been discovered in the tropics of Africa. "It was the first time this particular species had been discovered in the Northern Hemisphere," he says.

Bourland's discovery showed that *P. dibryophryis* is more widespread than previously thought. How remarkable that the same species of ciliate found in the rainforests of Africa could be swimming around in a mud puddle in Boise's Ann Morrison Park! Bourland smiles in the telling of his discovery.

So that's one value of a microbe hunter searching for new species more than 300 years after the original microbe hunters began documenting them. It helps us understand the uniqueness and range of ciliated protozoa on a global scale. "It gives us an idea of how things evolved on Earth," Bourland says. "How things are distributed—where they live, why they live where they

modern-day microbe hunter 27

Below:
A famous 1862 experiment with a swan-neck flask helped Pasteur demonstrate that food germs did not spontaneously generate and that microbes could be airborne in flecks of dust.

Institute Pasteur/Museum Science Group

live, how they contribute to the ecology of where they live."

Terrestrial ciliates are "enormously important for maintaining soil health," Bourland explains. "They transform nutrients in the soil, they eat bacteria and fungi, and control those things to some extent, and they provide food for other organisms, so they're part of the microbial food web. In that way, they're really important in uncultivated soils, agricultural soils and in aquatic environments."

But what's the deeper meaning? I ask Bourland. Why is it important to search for new ciliates? Why should anyone care?

"Because they're so remarkable," he says. "They are single-celled organisms, but they have all of the systems in place of much more complicated organisms. They have a way to encode their genetic information. They have a way to reproduce. They have a way to exchange genetic information sexually. They have a means of nourishing themselves—by different means. They can be carnivorous. They can eat bacteria and algae. They can absorb nutrients through their cell membrane and not eat anything at all. They have a way to move about. They have excretory systems for liquid and solid waste."

"And they have systems for defending themselves from other predators," he continues. "They have means of paralyzing their prey. So they are phenomenally complicated organisms, yet all of those systems are contained in one cell. From the standpoint of the organization of living things, they're really remarkable. They have in one cell all of the division of labor that our bodies require with its billions of different cells."

Ciliated protozoa contain two types of nuclei—a small diploid micronucleus that oversees reproductive functions and a large polyploid macronucleus that oversees general cell operations.

Left:
Mosquitoes host protozoa, viruses, and bacteria that kill more than a million people each year. Altering the genetic structure of the mosquito-transmitted microbes may help control the spread of disease.

Below:
Army physician Walter Reed (1851-1902) implicated the mosquito in the transmission of yellow fever. Reed's microbiology stemmed jungle mortality rates during the building of the Panama Canal.

The word *polyploid* means that the macronucleus contains more than two paired sets of chromosomes, compared with a microbe that is *diploid*, or contains a set of chromosomes from each parent. Ciliates reproduce asexually by various kinds of fission; each new cell contains a copy of the micronucleus and the macronucleus. They also reproduce sexually, exchanging micronuclear DNA in a process known as *conjugation*.

Bill Bourland was a curious kid growing up in Rockford, Illinois. His grandfather, who also was a physician, left behind an old brass German microscope. When 8-year-old Bill was poking around in his grandmother's attic, he found it and took it home. "In those days, all serious doctors had their own microscope. They had to do their own blood smears and lab work," Bourland says. "My grandfather's microscope was a very good microscope in its day." Bourland started using it immediately, looking at bacteria and blood smears—pricking his own finger for samples. "It really fascinated me," he says. "I set up my own microscopy lab with all kinds of materials—

modern-day microbe hunter 29

Science Source

more materials than I probably should have been able to have as a kid. And later, I got interested in protozoa and water samples. I thought it was amazing that you could see so many different protozoa in a teaspoon of water. And there would be dozens, if not hundreds of types of protozoa in a small jelly jar."

He remained interested in science through high school and earned a bachelor's degree in zoology from the University of Iowa.

While in college, Bourland learned about transmission electron microscopy, a technique that illuminates specimens by running a beam of electrons through an ultra-thin slide. But he had to put his passion for microscopy on hold while he went to medical school at the University of Iowa and completed his residency in general surgery at the University of Washington. He launched his surgery practice in Boise in 1986 and practiced medicine until 2008, when he started his work as a citizen scientist.

One day, Bourland got a wild hair and bought his first microscope, a Leica. It had bright field optics and phase contrast for being able to see different features of cells. "That was a big step up for me," he says. When he upgraded and added a feature called differential interference contrast (DIC), he could see the one-celled creatures in three-dimensional relief. That made it easier to identify different features of ciliated protozoa, apply stains to the specimens and take digital photos of the species.

"That really opened up a whole new world to me," he says, grinning. "It's like an astronomer getting a Hubble telescope. It's completely captivating. You could sit there for hours looking at one drop of water."

When he first inquired about adding images to the Micro*scope website, Bourland met David Patterson from the University of Sydney, Australia. "Paddy," as he's known to friends, was very helpful. He not only invited Bourland to contribute images, he visited him in Boise on several occasions. "He's

Left:
Freshwater *Tetrahymena thermophila* stained and magnified 3000x. Bottom: *Etoschophrya inornata* is one of four unique species of protozoa discovered by Bourland's team in his Boise State University lab.

Below:
A microbiologist grows bacteria in petri dishes. The spherical disk takes its name from Julius Richard Petri, who assisted the legendary bacteriologist Robert Koch of Berlin.

a great guy. He really motivated me to go to the next level," Bourland says. "We looked at samples together, all sorts of things. Algae, bacteria, amoeba and flagellates, and Paddy was instrumental in helping me refine my study to ciliated protozoa. I owe him a great debt."

Bourland also met Wilhelm Foissner, the world's preeminent ciliatologist, who teaches at the University of Salzburg, Austria. Their relationship started through an innocent e-mail with a question. Bourland expected him to say, "Don't

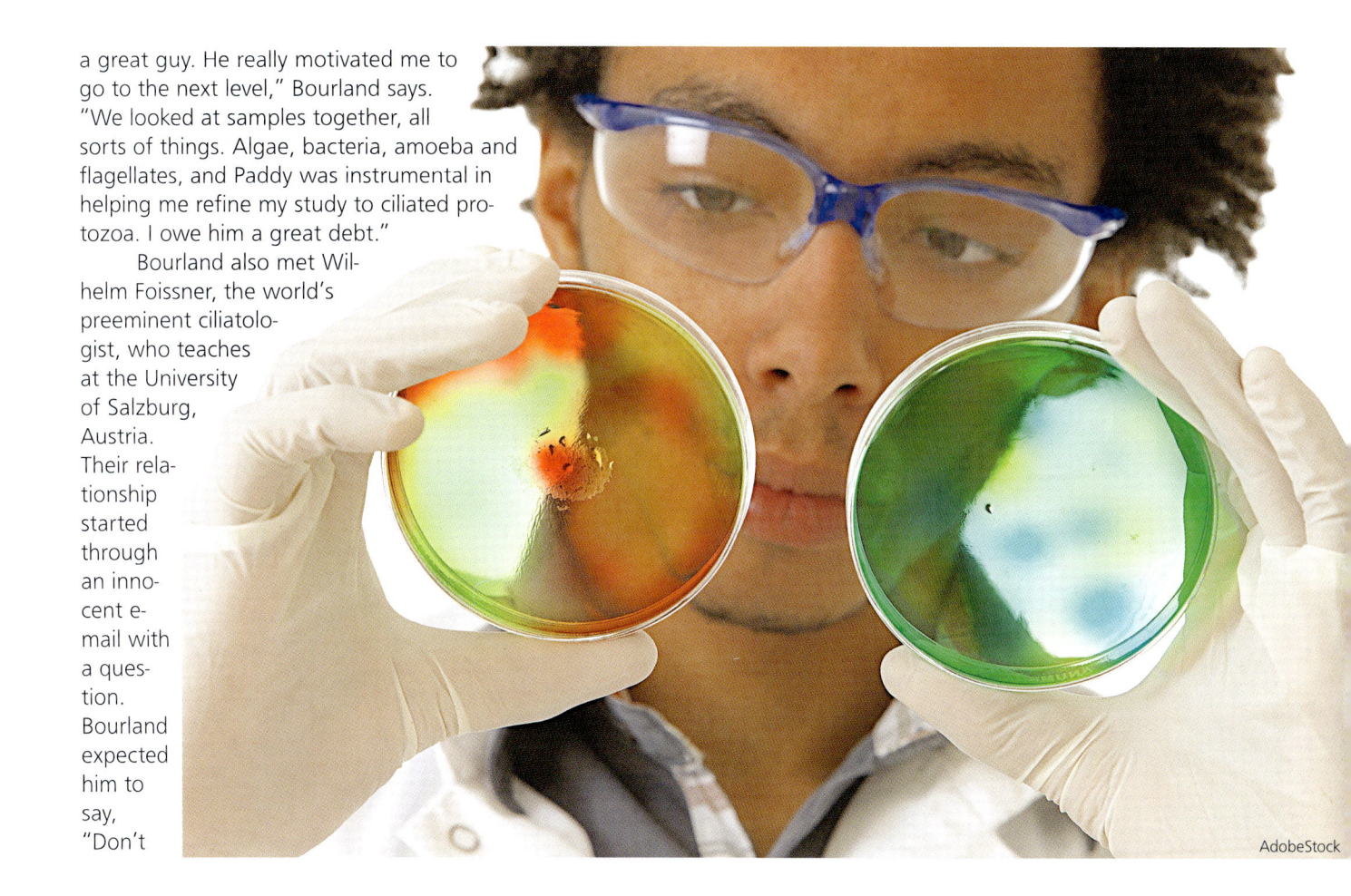

modern-day microbe hunter 31

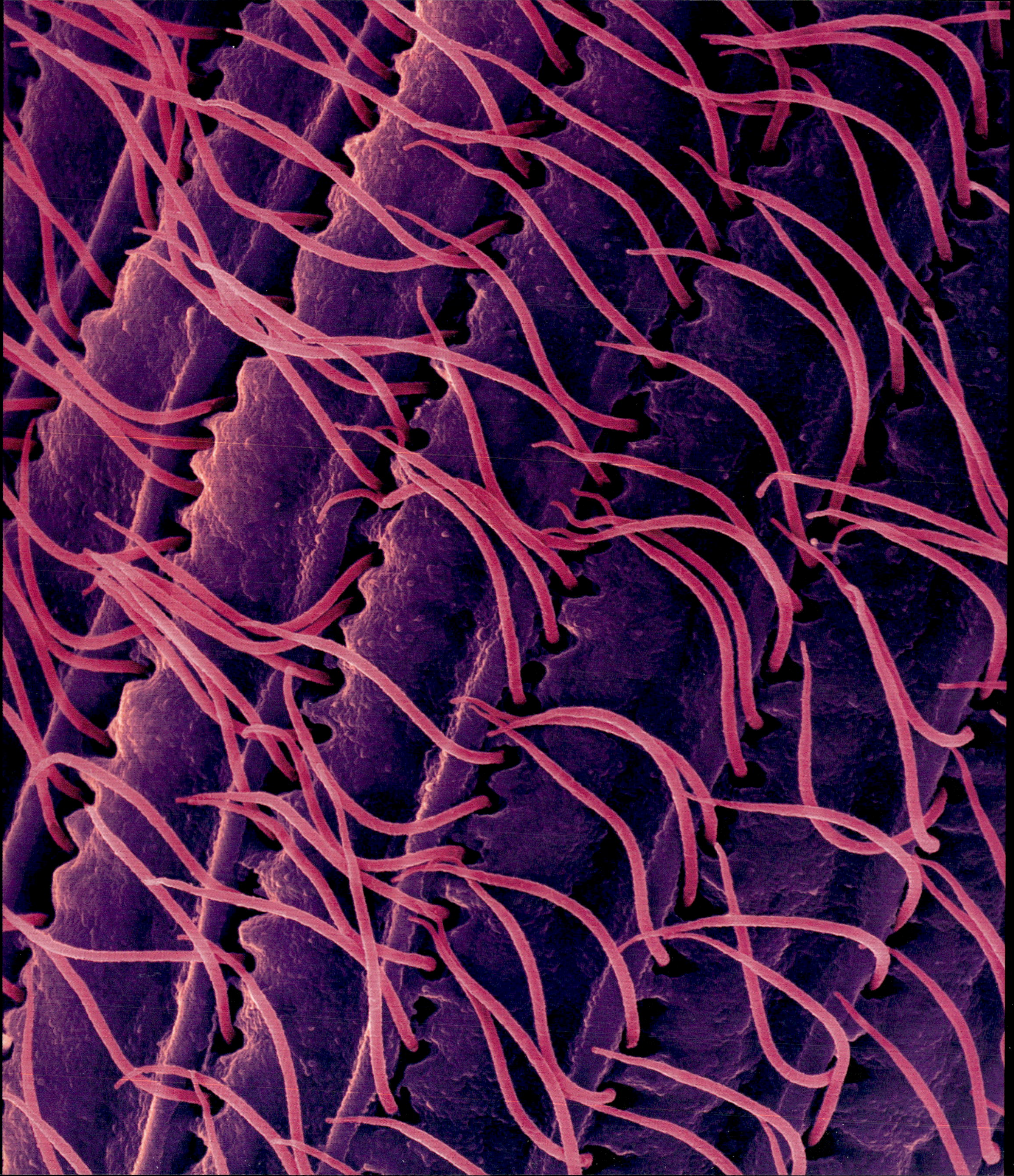
Dennis Kunkel Microscopy

Left: Cilia of *Tetrahymena thermophila*, stained pink and electronically scanned. Seven sexes give the ciliate a range of mating options that help it cope with extreme environments.

Below: Lazzaro Spallanzani (1729-1799) theorized that microbes move through the air. Joseph Lister (1827-1912) pioneered germ theory and antiseptic surgery.

bother me, you're not one of us." But actually, Foissner was very helpful, immediately. "He was very gracious," Bourland says. "He wrote a long e-mail back to me. He showed me how to be more detailed in my observations and what to look for."

Foissner even invited Bourland to visit Austria and see his lab, where he taught the citizen scientist more tricks of the trade. "He had learned all of the methods that are needed for studying these ciliates in detail—silver-staining methods that had been around since the early 1900s but are very difficult to learn, very complicated," Bourland says. "It was very useful for me to learn these methods. He's probably the best in the world, so it was like studying at the feet of the master."

While visiting Foissner, Bourland met another leading citizen scientist, Martin Kreutz of Germany. "He takes beautiful pictures of protozoa using the DIC technique," Bourland says. And he met Peter Vďacný, a leading ciliatologist from Comenius University in Bratislava. On a single trip to western Europe, Bourland had been invited into the inner fold of the top protistologists in the world. After these visits, he was totally hooked, and he came home to Boise freshly energized to practice what he'd learned.

In August 2012, Bourland presented a poster at an international meeting of protistologists in Oslo, Norway. The poster is populated with high-quality color photographs of *Balantidium pellucidum*, a freshwater ciliate that Bourland characterizes in a resting cyst form. When this particular ciliate is exposed to dry conditions, its survival mechanisms kick in and it turns into a cyst, waiting for the next rain. "If they have no food, they'll ball up and go into this resting state and remain that way for some time," Bourland says. "It's called crytobiosis. They can persist in that form for years until moisture returns. When rain falls on arid soil, they sense the change and dissolve the cyst wall."

modern-day microbe hunter 33

The lesson here, Bourland says, and the question that remains is to understand how "cells preserve themselves for that period of time. That has all sorts of implications for cell preservation, organ storage and how a microorganism can survive extreme environments."

Bourland isn't worried about running out of things to do when it comes to discovering new ciliated protozoa and understanding how they tick. "There are literally thousands of ciliates yet to be discovered," he says. "Some experts suggest that there are more to be discovered than have been discovered so far. It's a matter of how hard people are looking for them or how carefully."

Other scientific fields such as biochemistry, molecular methods, DNA and genetics have shifted some interest away from classical alpha taxonomy—the discovery and description of new species, Bourland says. "A lot of the techniques that are needed to study things are being lost—like the silver-impregnation techniques—and people

Left:
Cryptobiosis allows microbes to retract and shut down in a state of suspended animation. Pictured: The microscopic water bear, when suspended in cryptobiosis, can survive even in space.

Below:
The skeletal beauty of zooplankton protozoa captivated Ernst Haeckel, a German zoologist and illustrator. Pictured: *Circogonia icosahedra* from Haeckel's *Kunstformen der Natur*, 1899.

need to learn it almost like an art. It's important for it to be passed along to new generations of science students and scientists so it's not lost over time."

Because of the intrinsic value of his work as a citizen scientist, Bourland is now a research scientist in the Department of Biological Sciences at Boise State University, teaching as an adjunct graduate research faculty member and working in the lab, using DNA, electron microscopy and classical staining techniques to identify and describe new species of ciliates. Bourland is working with his first graduate student, Laura Wendell, passing on his passion and knowledge, as she investigates the effects of Idaho's frequent wildfires on the amazingly resilient ciliates in sagebrush steppe soil.

It seems that our world is better off when we have self-driven inquisitive people like Bourland hunting for new microbes as a daily pursuit, while sharing the information with a close-knit group of ciliate experts worldwide. His discoveries will have a lasting value for science and our understanding of the world. As Margaret Mead said, "Never doubt that a small group of thoughtful, committed citizens can change the world. Indeed, it is the only thing that ever has."

Life in a Drop of W

More than an eighth of the world's population drinks water with dangerous microbes. Waterborne diseases kill more than 3.6 million people each year.

What dangers lurk in a nearby drop of untreated water? Start by filling a few jars with water from a nearby pond, river or puddle. Put a small amount of cut grass or hay in each jar. Put lids on the jars and allow them to sit for at least one week. (Caution: Wash hands thoroughly after working with hay. Don't swallow the water or allow it to get in your eyes.) Place jars on a tray. Open cautiously. The water will smell terrible and could overflow.

Under a microscope, using sterile slides, compare hay/grass water droplets to tap water from your faucet. More than 250 million bacteria can inhabit a millimeter of untreated water. From place to place you will find considerable

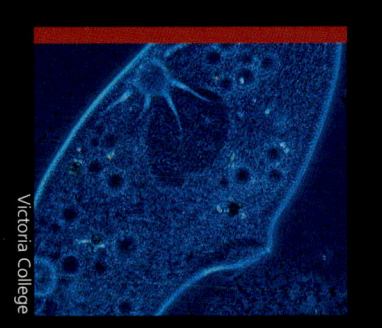

Ideaventions

Paramacia — Victoria College

Rotifer — Jason LeBlanc/Lamar University

Stentors — microscopy-UK

Amoebae — environmentalleverage.com

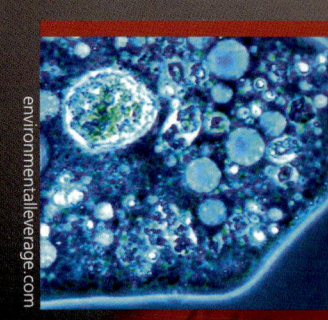

variation. Paramecia, rotifers, stentors, euglenae and amoebae are common in Idaho puddles. Note that certain protists (such as hydras and certain rotifers) are stationary and will be found anchored on some kind of debris.

Nine of ten Idahoans rely on wells deep enough to be safe from the dangerous diseases. In the developing world, nevertheless, untreated water remains one of the world's deadliest killers. Cholera in Haiti has infected more than 400,000 since an earthquake rocked the island in 2010.

Paul Mood/Inspiring Wallpapers

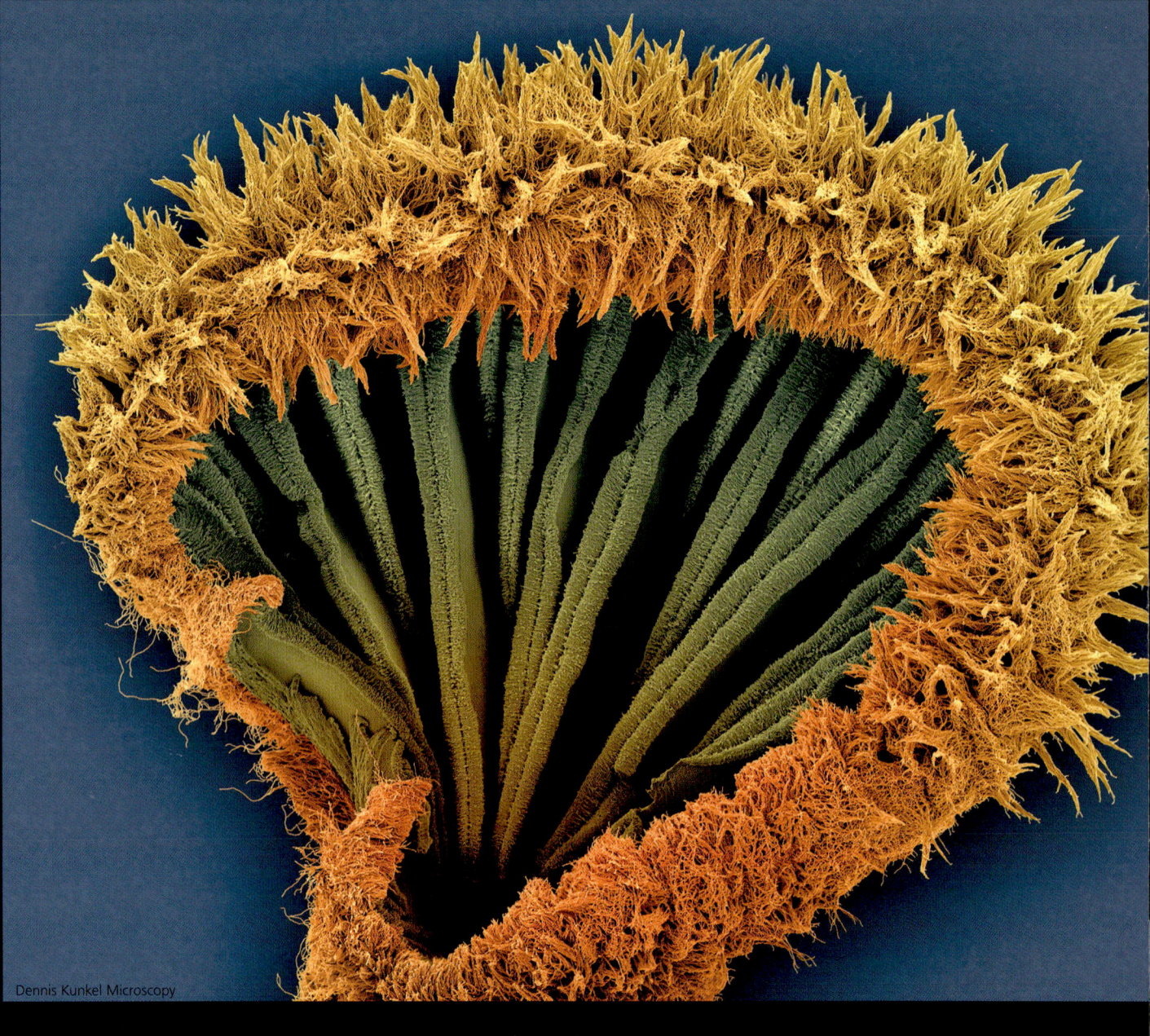

Bracket Fungus Basidiocarp

Dennis Kunkel Microscopy

wondrous mycelia
and the hunger for natural foods

2

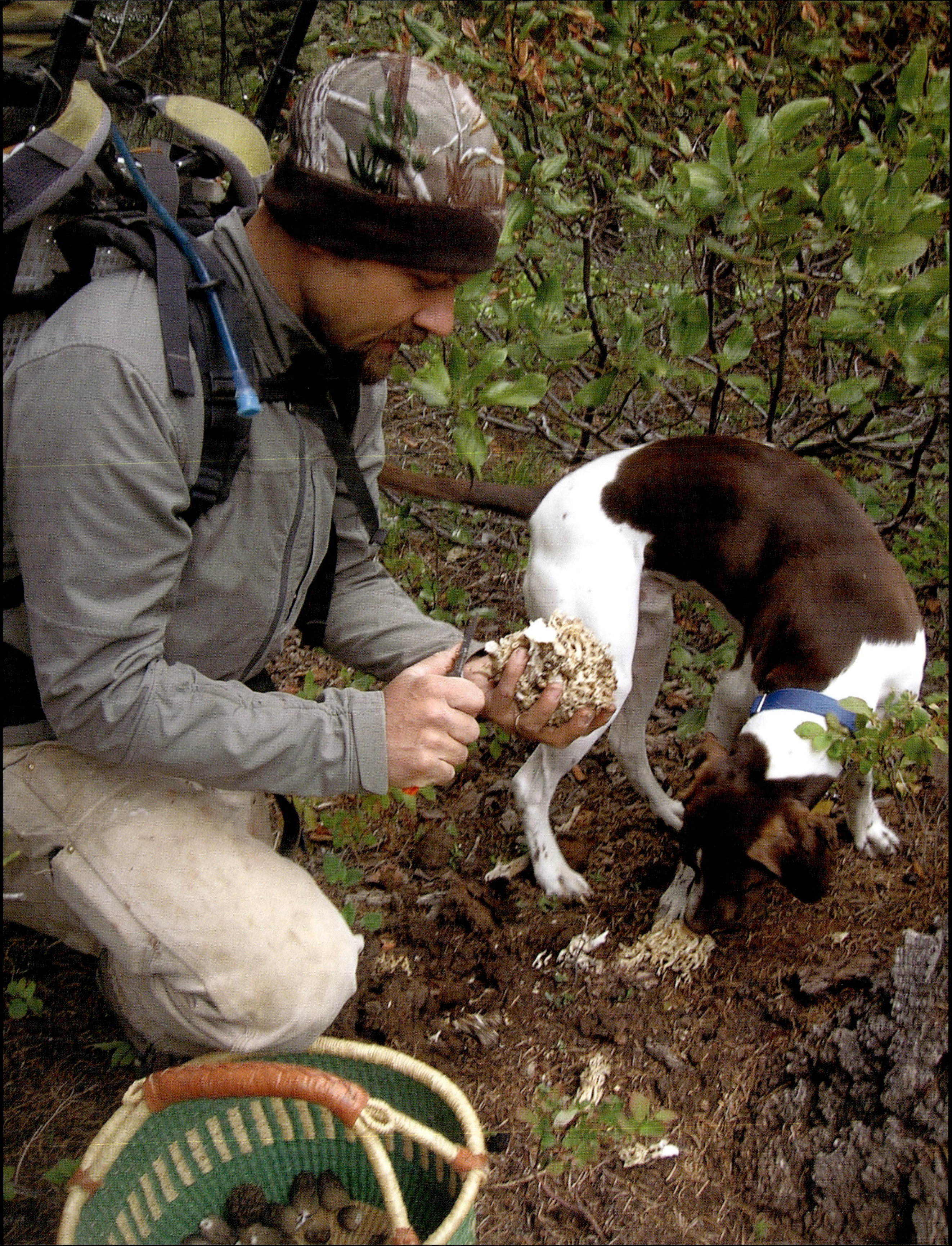

Steve Stuebner

Left:
Chris Florence of Sweet Valley Family Farms forages for edible mushrooms in the forests of the Pacific Northwest. With a keen eye for delectable fungi, he searches shady places near dead or dying trees.

Previous:
Fungi have cells with porous walls that permit the flow of fluids and proteins. Pictured: Bracket fungus, or shelf fungus, protrudes from the trunks of trees like brackets holding shelves.

On a cloudy Sunday in late May, Chris Florence of Sweet Valley Family Farms takes me foraging in the Boise National Forest. We drive separate vehicles to Smiths Ferry, and I can barely keep up with him in his 1997 green Subaru Outback. We're out early, there's no traffic, and he zips along on Idaho Highway 55 at high speed. This is a guy who travels 45,000 miles a year, foraging for natural foods in remote mountain locations. I try to keep him in sight.

We turn onto U.S. Forest Service roads in Smiths Ferry and pass numerous campers on our way to Florence's favorite foraging spot in this neck of the woods. He knows the country well—he's been hunting for morels and other edible natural foods in the forests of Idaho for 14 years. He keeps track of his favorite spots and returns to them frequently to forage for natural food.

We reach the area where Florence wants to hike up the mountain and look for morel mushrooms, maybe some nettles and perhaps some calf brain mushrooms. There is no trail. Florence has four, white rectangular baskets that he ties onto a backpack frame, and he carries a green wicker picnic basket by hand. This is where he will place any morels that we find on the hike. He's wearing a camo stocking cap (which he wears when he's bow-hunting for elk or deer), Carhartt pants, hiking boots and a raincoat. This is a man who's at home in the woods.

Florence hikes directly up the mountain, bushwhacking through the brush. It's a moderate climb along the nose of a ridge. He finds some snow mushrooms growing in the dirt. "They're kind of bland, so I don't collect them," he says. He sees a lump of dirt, takes out his pocketknife and brushes away the soil to reveal a coral mushroom, aptly named because they look similar to coral that you might see while snorkeling in the ocean. It's like a small shrub with lots of branches. "They're edible, but they're hard to clean," he says, passing on them as well.

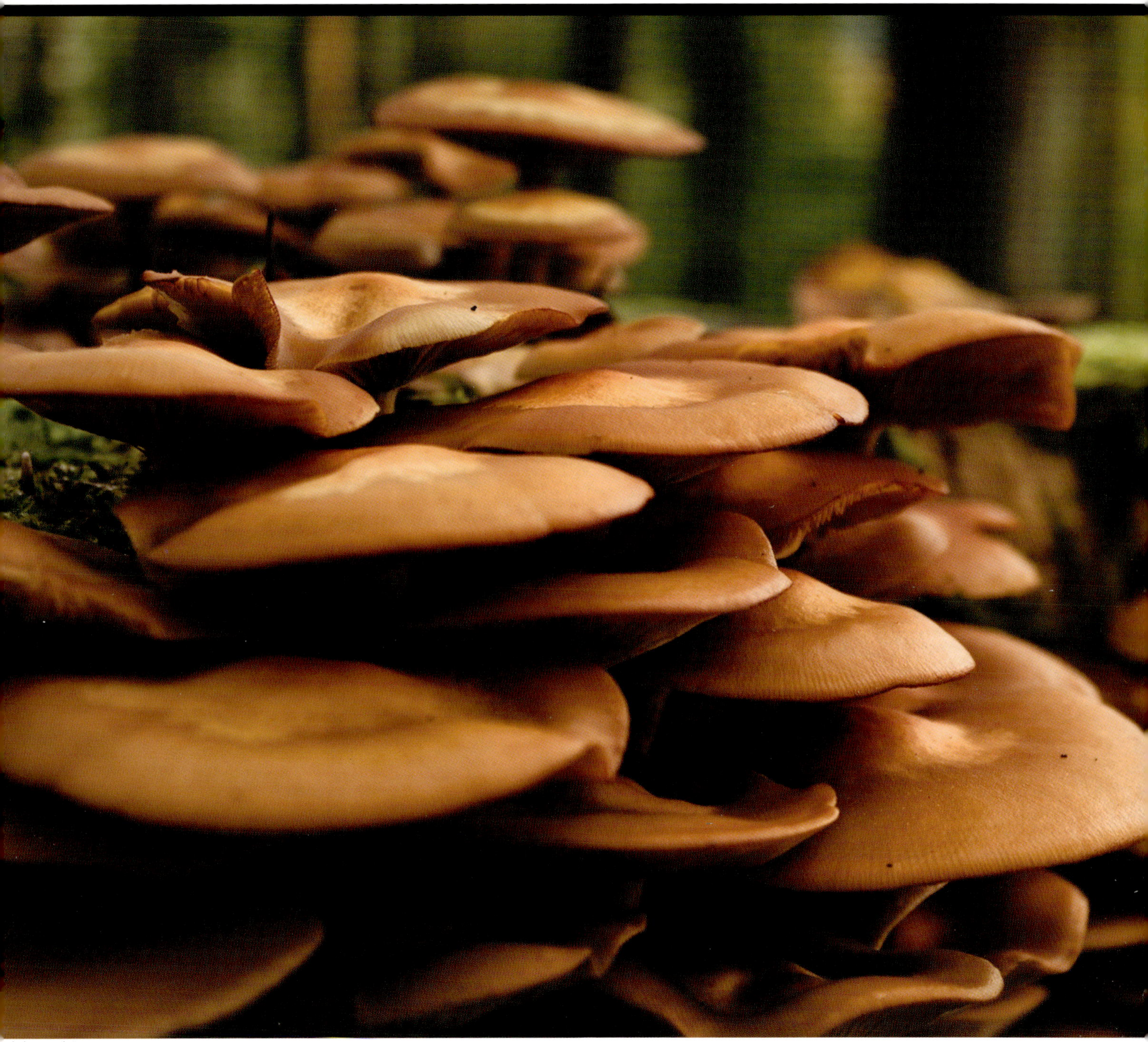

Left:
Mushrooms blanket a forest. Their vast substrate networks of mycelia sprawl for miles under the surface of the soil.

Below:
In April or May, when trilliums bloom purple in woodlands, it's prime time for picking morels. Bottom: puffs of white mycelium spreading from a tree stump.

I pepper him with questions about how he finds morels. I've been a citizen morel hunter for years, but I've never really seen a pattern to where the mushrooms will grow. I want to learn more. Do they prefer growing next to pine trees or fir trees? "I see more of them closer to fir than to pine," he says. "They colonize with trees, shrubs, even cheatgrass. Sometimes I find them more on a particular slope aspect or near various types of vegetation. It really can be anything."

More than anything else, Florence

wondrous mycelia 43

says the timing of when morels are popping is key, and tracking where they are emerging is another watch point. Right now, they're popping up in the 5,300- to 6,000-foot elevation zone. Legend has it, Florence says, that if the purple flowers of trillium are blooming, then it's time to pick morels. We happened to see trilliums flowering everywhere around us as we hunted for morels that day.

Hunting for morels can be both exciting and frustrating. Morels come from the genus *Morchella*. They're treasured around the world because of their delicious taste and nutritious value. They're shaped like a miniature Christmas tree. The heads are made up of an intricate architecture of vertical honeycombs, making them easy to identify. Clyde Christensen, author of *Common Edible Mushrooms*, refers to morels as one of the "Infallible Four" in terms of their looks. They're pretty darn safe to collect; plus, they don't have any close cousins that look the same but are poisonous.

For Florence, collecting morels is not a hobby, it's a business. He's the chief forager for Sweet Valley Family Farms, a natural foods wholesaler based in the Boise Valley. Morels can fetch in the neighborhood of $20 a pound at farmers markets in Boise, and Florence sells premium edible mushrooms to Boise and Sun Valley restaurants.

Florence continues hiking up the ridge and whispers, "This is a really shroomy area." Within minutes, he reaches for his pocketknife and snips off several small morels at the stems just above the surface of the ground. He's

Left:
Elusive morels are honeycombed with networks of ridges. Delicious when sautéed with butter and garlic, they fetch high prices at farmers markets.

Below:
Sweet Valley's farm beneath amber foothills is propane-free with geothermal heating and fuel recycled from restaurant kitchen grease. Pictured: Florence with seedling tomatoes.

got eagle eyes for morels. I'm staring at the ground as we walk, and all I can see are branches, downed trees, shrubs and wildflowers. Florence finds a patch of blond morels and snips them off with his knife. I've never seen a blond morel before. "They have a different, distinct flavor," he says.

It starts to rain lightly, but we keep pushing up the slope. Florence doesn't find any large patches of morels, but he keeps finding large and small individual morels growing underneath alder bushes and out in the open. He digs into the dirt to show me the white-streaked layer of mycelia growing just an inch or so below the surface. "There's a mycorrhizal layer that grows in the soil throughout the forest," he explains. "The mycelia have a relationship with the plants and trees, but we only know a little bit about that. The mycelia grow around the root hair and share nutrients with plants."

As veteran morel hunters know, the mushrooms often sprout in areas that have been disturbed in some way. They often grow in prolific numbers in the spring after a forest fire has occurred the previous year, or they'll sprout in areas where wildlife or humans have disrupted the soil in some way, such as on

Dennis Kunkel Microscopy

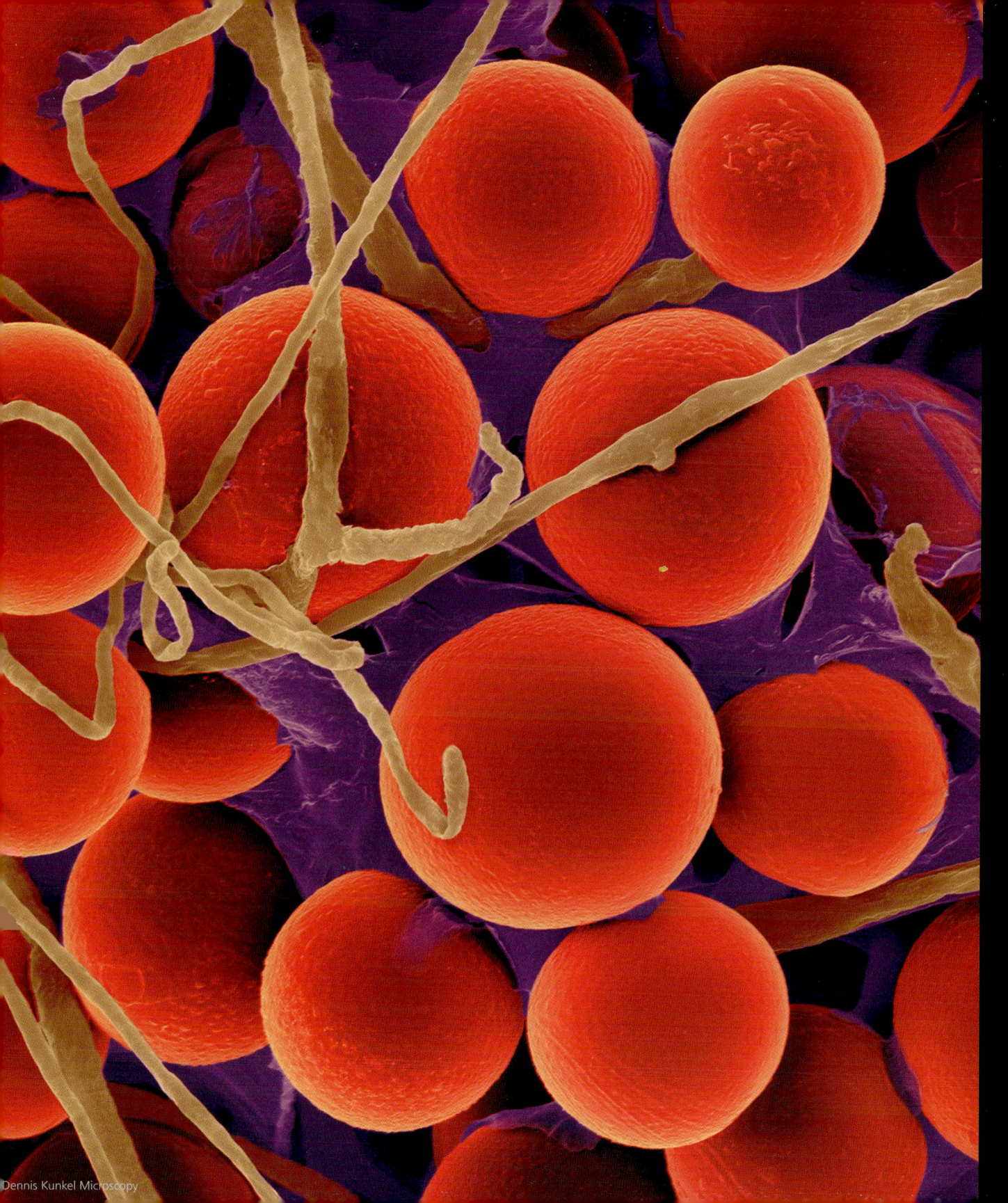

Left:
The sexual spores of *Aspergillus nidulans* can grow in the soil or in household molds. A single mushroom can produce billions of spores.

Below:
Yellowish morels with elongated heads are the most common of the edible species. Lopsided reddish brown "false morels" can be deadly toxic.

logging skid trails. "They like areas of disturbance—that's what causes them to bloom," Florence says. "Fire can destroy the root structure of the soil, depending on how hot it burns, and that's when the mycorrhizal layer puts all of its energy into kicking out as many spores as possible to survive."

The manifestation of this energy is a mushroom emerging from the ground, which, by its very nature, will kick out more spores and build a colony of mushrooms as weather conditions allow. After a few days, the magic moment may pass as the temperatures heat up and the morels dry up, and it's time to climb to higher elevations to find them. This is why Florence is constantly on the run.

Venture into the forest and dig down a few inches to see the furry, white layer of mycelia just an inch or so below the surface. Pick up a piece of dead wood from the forest floor and peer underneath, and you'll see how the mycelia decompose the woody material and convert it into soil. When mycelium is viewed under the microscope, it looks like a web of life, an intricate network of cells that nourishes the soil and the plant life in the forest.

wildculture.com

Below:
The life cycle of fungi. In mushrooms, the gills of the fruiting body release spores, producing the hyphal cells that network with the mycelia and cause new mushrooms to form.

Science Source

"There are more species of fungi, bacteria and protozoa in a single scoop of soil than there are species of plants and vertebrate animals in all of North America," writes Paul Stamets, author of *Mycelium Running: How Mushrooms Can Save the World*. "And of these, fungi are the grand recyclers of our planet, the mycomagicians disassembling large organic molecules into simpler forms, which in turn nourish other members of the ecological community. Fungi are the interface organisms between life and death."

Stamets compares the web of life provided by mycelia with the network of stars and planets in the universe or even computer networks that span the globe. "Fungi are keystone species that create ever-thickening layers of soil, which allow future plant and animal generations to flourish. Without fungi, all ecosystems would fail," Stamets writes.

Historically, mycelia have been part of the Earth's

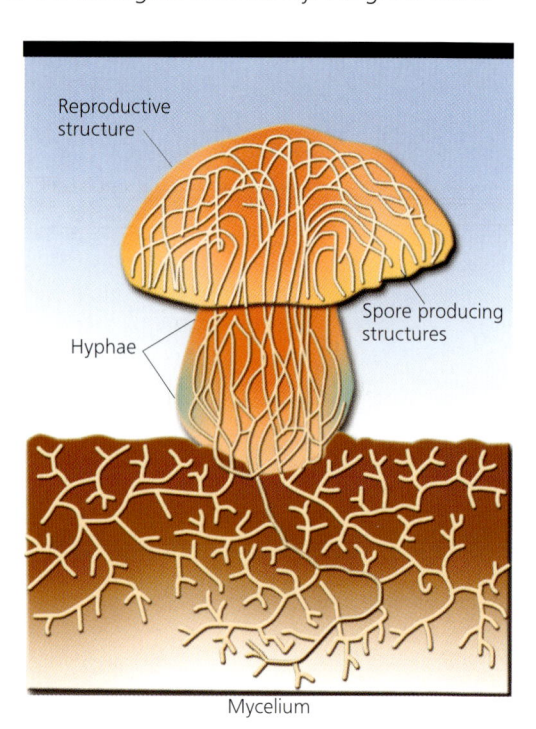

Reproductive structure
Spore producing structures
Hyphae
Mycelium

Below:
Mycelia are tangled with threadlike tubular hyphae too small to be seen by the eye. Mycologists have catalogued more than 13,000 species.

ecosystem for more than a billion years. They've survived several major catastrophes, including an event spurred by a meteorite 250 million years ago that wiped out 90 percent of the Earth's species, Stamets says, and again, 185 million years later, when a meteorite struck the Earth and wiped out the dinosaurs and many other species. Mycelia survived by partnering with plants and sending out spores to perpetuate the species. The oldest-known cap and stem mushroom dates to 92 million to 94 million years ago, Stamets says.

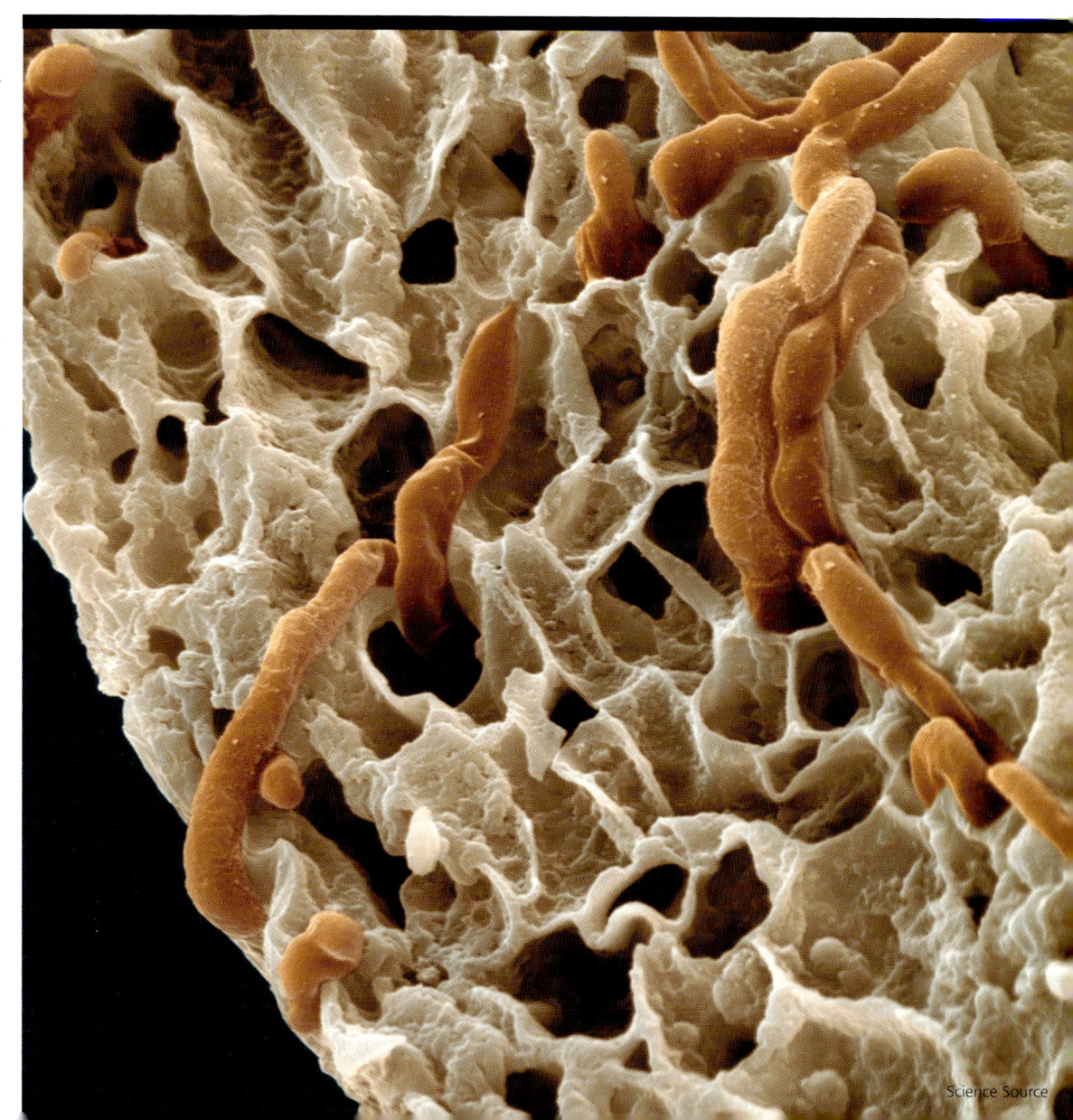

Science Source

David M. Cobb Photography

Left:
Gills rib the poison pie mushroom's delicate underside. An important foraging clue to species identification, gills disperse the spores.

Below:
Mushrooms are the fruit of the fungus that disperses the spores. The spores grow the mycelium that give rise to the "button" or "cap."

In general, the life cycle of a mushroom begins when a mature mushroom launches spores from the basidia, the hairlike follicles on the underside of a mushroom cap. Most mushroom species cast away four-spored basidia, which are jettisoned in pairs with enough force to throw them inches away from the mushroom.

The spores tend to fall near their parent mushroom, and trails of spores can be seen wafting in the air as well. Stamets points out that insects and mammals help with the distribution of spores. Drawn by the scent of mushrooms, insects use them as a home for their larvae and carry spores with them when they leave the nest. Many mammals eat mushrooms for food, and some of the spores survive the digestive process and get dispersed via animal waste. For example, a researcher at Oregon State University discovered that because of animal waste, the range of subterranean truffle mushrooms is more widely dispersed than it would be otherwise.

Large *Ganoderma applanatum* mushrooms, the pancake-like shrooms that grow out of the sides of trees, liberate an estimated 30 billion spores a day and more than 5 trillion a year, Stamets says. The number of spores sent out depends on the mushroom species. The timing and duration of spore release depends on moisture, temperatures, forest habitat and other factors.

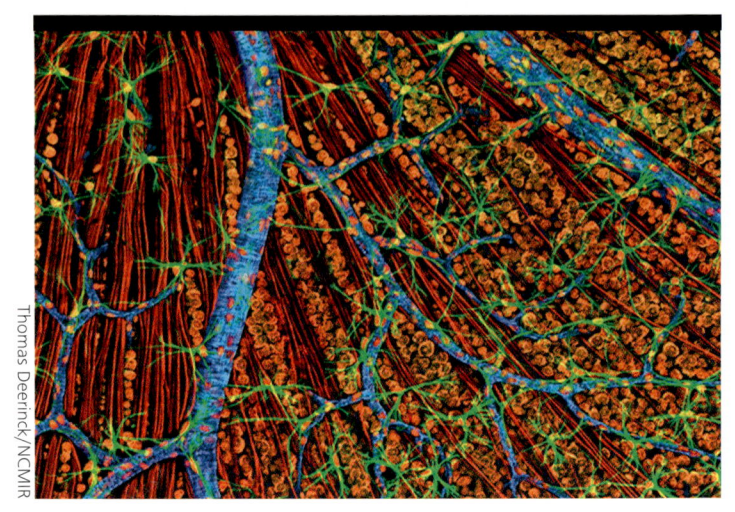

Germination begins in the dimpled depression on the spore. In the first few minutes, it looks like a seed is sprouting, Stamets says. The sprout-like hyphae of the basidiospores divide via the process of mitosis, cell division by which the nucleus divides, typically consisting of four stages (prophase, metaphase, anaphase and telophase), resulting in two new nuclei, each of which contains a complete copy of the parental chromosomes.

Next, the hyphae mate from two compatible spores and fuse to form one mycelium. The resulting cellular network, called a dikaryon, is "invigorated" to reproduce further, producing more fertile mushrooms with spore-bearing ability.

In the life cycle of morel mushrooms, a separate step occurs after the hyphae have mated and formed mycelia. During the germination of a new mushroom, there is a step called sclerotium, a large structure (1 to 5 cm in diameter) composed of large cells with thick cell walls that allow morels to survive adverse natural conditions, such as the extreme heat of forest fires.

Because of the ubiquitous nature of the mycelia and mushrooms in the forest, each species is capable of perpetuating itself at a grand scale. Dozens of different mushroom

Left:
Paul Stamets enlists fungi in the fight against human pandemics. In his book *Mycelium Running* (2005), he compares mycelia to intricate natural systems such as (bottom) the fibers of an optic nerve.

Below:
Poisonous mushrooms poster from France, about 1890. The fly amanita (far left) is a hallucinogenic that was once, in Christian literature, an allegory for Jesus Christ.

species pop up in the forest in a micro-environment that's perfect for them, and they are constantly sending out spores to spawn new generations while conditions are ripe, while the mycelium layer underneath the surface maintains its web of life throughout the year.

Stamets puts it into perspective: "In a gram of myceliated soil, more than a mile of cells form. In a cubic inch, more than 8 miles of cells form." In a photo of his hiking boot in *Mycelium Running*, Stamets says that footprint covers 300 miles of mycelium underfoot. Hence, from a mycelium's point of reference, a "journey of 10,000 miles is only 33+ footsteps."

Through 14 years of foraging in Idaho and the Pacific Northwest, Florence has spent countless hours roaming the woods, learning where to find the most prized edible mushrooms. Beyond hunting for six species of morels, he forages for boletes (porcini and cèpes are his all-time favorites), calf brains, coral mushrooms, cauliflower mushrooms, chanterelles and hedgehogs.

He also collects edible foods—nettles, miner's lettuce, wild watercress, wild mint, elderflowers, elderberries, huckleberries, wild

pinterest.com/virginiadempsey

wondrous mycelia 53

Michael Hoffman/Adele Thomsen

Florentine Codex

fruit (apples, pears and crab apples) and cottonwood buds. Sweet Valley Family Farms sells all of those natural foods at the Boise farmers markets on Saturdays and sells them wholesale to restaurants such as Bittercreek and Red Feather in Boise; CK's Real Food in Hailey; Globus and Ketchum Grill in Ketchum; and Atkinsons' Markets in the Wood River Valley.

Given how these natural food products are perishable, imagine how quickly Florence must travel (with the air conditioner cranking) from remote mountain locations in his Subaru to the home base to deliver fresh product. "I drive over 45,000 miles a year," he says with a grin. "That includes miles for foraging as well as deliveries."

Florence doesn't do it all single-handedly. He has developed relationships over the years with many foragers like himself and often buys from other foragers when he doesn't have time to forage for certain products himself. "I used to forage 100 percent of what I sold, and then we got too big," he says. "I work with dozens of pickers and wholesale brokers and buy from them," he says. "I've made a lot of friends picking over the years."

When he's out foraging, Florence thinks a lot about early hunters and gatherers. "I think about that all the time," he says. "Gathering wild food is challenging. It takes a lot of knowledge and experience. I wonder what hunting and gathering societies were like. They had to be very efficient. They had to follow the seasons. They traveled hundreds of miles by foot or horseback. The knowledge they had was phenomenal. They had to be very educated because, after all, they had to find the food they needed. It was life or death! I'm really inspired by all of that."

Left:

Early Christians featured mushrooms in spiritual art. Pictured: mushroom trees in the stained glass of Chartres Cathedral, 1235; Aztec mushroom feast from the *Florentine Codex*, 1569.

Below:

Fungi range from single-celled organisms to mile-long mycelia in a kingdom suspended between microbiotics and macro multicellular life. Pictured: *Omphalotus olivascens*, a forest dweller.

wondrous mycelia

commons.wikimedia.org

Left:
Toadstools commonly refer to most any mushroom-like fungi that are toxic even when cooked. Pictured: white-spotted *Amanita muscaria*, the quintessential toadstool.

Below:
Hyphae are threadlike tubular cells that weave through the mycelium. Pictured: pastel-stained hyphae of *Penicillium*; branching hyphae of mold fungus.

Florence grew up in Meridian, Idaho. He started big game hunting and bird hunting as a young teenager. He and his friends would wake up early to go duck hunting along the Boise River in the fall and still make it to school on time. Over the years, he switched from rifle hunting for big game to bow hunting. He learned how to call in a bull elk—mimicking the high-pitched bugle used by a bull to call in the cows for mating—so that he could shoot it at close range. Killing and eating wild game birds, waterfowl and big game animals were his first venture into eating wild, natural foods.

After high school, Florence took a liking to cooking. He trained as a chef at the California Culinary Academy and worked at a San Francisco restaurant. A friend took him into the Santa Cruz Mountains to look for edible mushrooms for the restaurant. "We found three mushrooms that day, and it was like finding a golden egg," Florence says. "I fell in love with natural ingredients and being a part of the process of finding those ingredients from start to finish. I started learning about processed foods and became concerned about what kinds of foods and farm chemicals that I put into my body."

Ultimately, Florence found the idea of gathering natural foods in the forest as a way to make money quite appealing. He'd been roaming the woods most of his life. "It's like a dream job, right? Being able to be out in the woods and making a living is something I love to do. I just go out and do my thing, and I don't have to answer to anyone."

It's a sizzling hot Saturday in late June, and Sweet Valley Organics (now defunct) is hosting a group of about 35 people on a tour of two organic farms in the Emmett area. The tour is sponsored by the Treasure Valley Food Coalition and the Northwest Center for Alternatives to Pesticides. The food coalition supports the

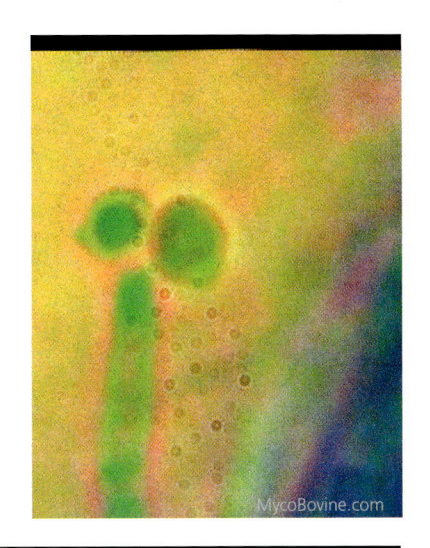

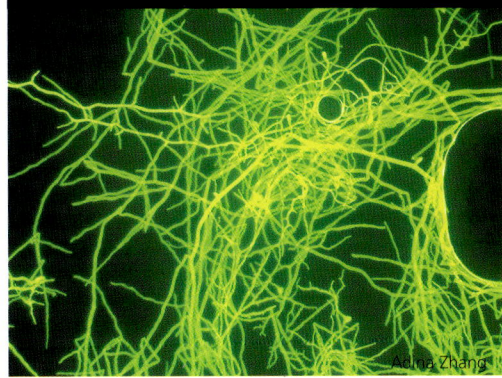

wondrous mycelia 57

cultivation of locally grown, organic food as a way to promote healthy lifestyles and a healthy community.

Janie Burns, owner of Meadowlark Farm in Nampa, explains how the farming system in the United States has changed over time. In the old days, she says, family farmers would raise food for their families and sell crops on the open market. They'd raise chickens, pigs, cattle, goats and dairy cows, and they'd plant a garden. Nowadays, a lot of farms are owned by corporations, and they grow large quantities of a few crops for the nation—or the world. "We're shipping out the products we grow, and we're shipping in the food we eat," Burns says.

That said, a recent economic report from the University of Idaho Extension Service indicates that Idaho's farms and ranches generate $21 billion in total sales, pay $4.2 billion in wages, contribute $8.4 billion in gross state product and support 156,599 jobs. Idaho's farm base is one of the most diverse in the nation, with a dizzying array of products, from grains like wheat and barley in the Palouse to sugar beets, potatoes, dairy, beef and corn in southern Idaho. The export market from Idaho farms is valued at $8.9 billion, according to the University of Idaho report.

As the tour arrives to the Sweet Valley Organics Farm, the hot sun is getting lower in the sky—much to the relief of the group—and it casts a golden glow on the fields of heirloom tomatoes growing up rapidly on the north side of the farm. Sweet Valley Organics has 27.5 acres of farm ground, a greenhouse heated with geothermal heat, where they start the vegetable plants in the winter, and an open-pasture organic chicken operation. The food sold from the farm supports Florence's young family and two other families. A sign at the entrance says, "Welcome: Dirt Poor, Soil Rich."

"Farming for us is more than making a living," Florence says. "We want to improve the land so we can leave it in better shape than when we got

Below:
The jack-o'-lantern (*Omphalotus olearius*) is toxic. The pumpkin-orange funnel-shaped fungus is easily confused with the edible chanterelle.

Dennis Kunkel Microscopy

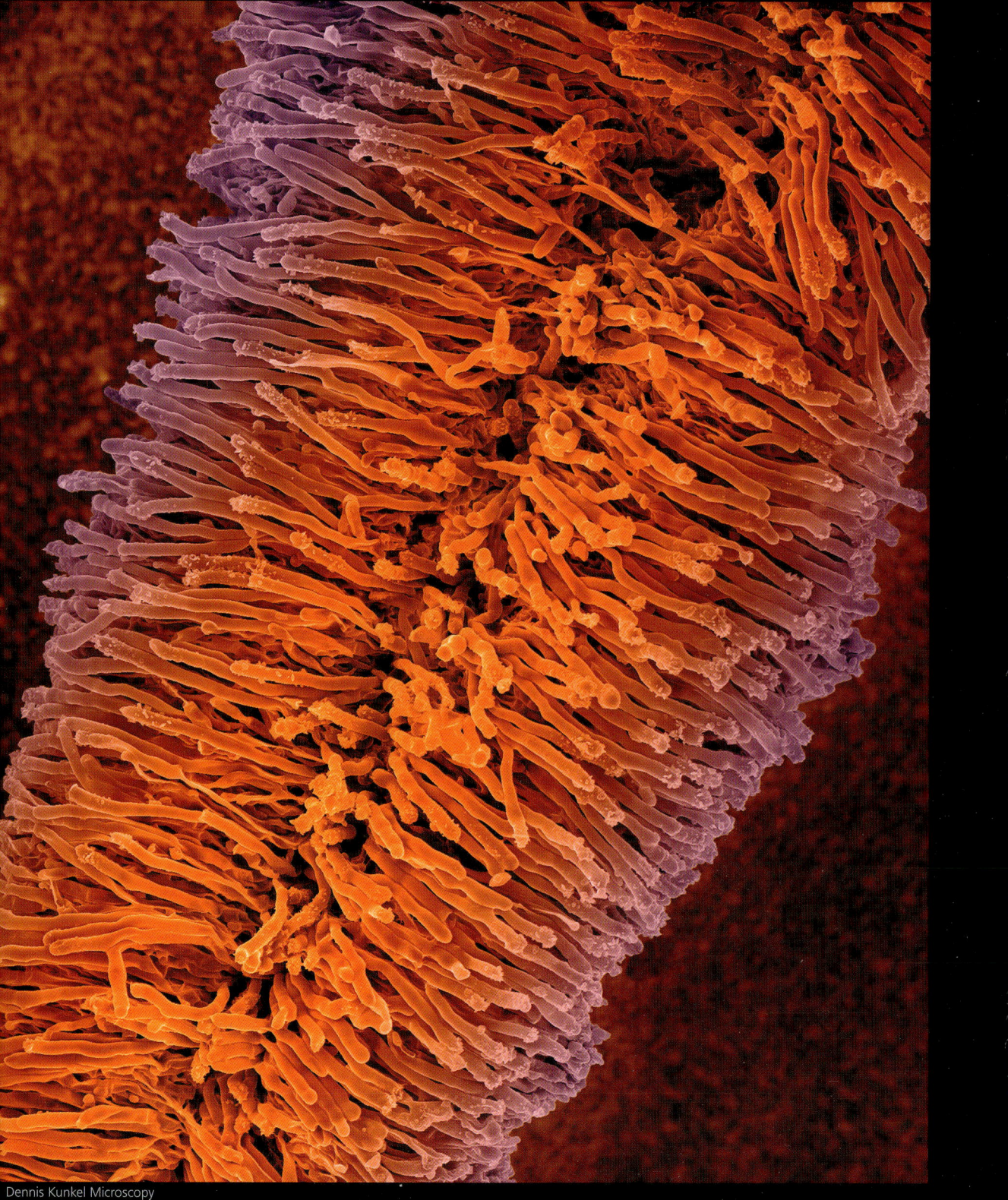

Left:
Hydnoid fungi, or tooth fungi, use spine-like toothy projections to disperse microscopic spores. Pictured: underside of bracket fungus with spore-bearing spines.

Below:
Sweet Valley practices vermiculture. Worms compost organic matter for organic fertilizers that return nutrients to the soil. Pictured: *Eisenia fetida*, commonly used in vermiculture.

here. We're trying to establish something that's sustainable that will last for generations. And we're really trying to push organic food in the valley."

Sweet Valley Organics started in 2008, just as an emerging citizen movement that embraced locally grown food was taking shape in southwest Idaho. Sweet Valley Organics grows heirloom tomatoes, squash, peppers and basil, and they have 800 laying chicken hens. They gather manure from their horses and livestock and work that into the topsoil for fertilizer, and they do vermicomposting, a method of composting using earthworms, which produces nutrient-rich organic fertilizer. They apply the compost and manure on the soil with a manure-spreader that dates to the 1950s. "It doesn't look like much, but it works like a dream," says Geoff Neyman, who is in charge of sustainable practices at the farm.

Pairing the wild, natural food foraging business with the organic farm makes a lot of sense, says Florence, who now sells and distributes organic food from about 15 farms. "The two go really well together," he says. "The same people who are looking for specialty produce are looking for wild mushrooms as well. There's a lot of overlap there."

Interest in Sweet Valley's products has been strong, Florence says, and the farmers markets in Idaho are growing as well, reflecting the rising interest by Gem State residents to buy locally grown food. The Boise Farmers Market in downtown Boise, for instance, had 30 to 40 food vendors in 2013, and now they have 60 to 70 vendors, including Sweet Valley Family Farms. Statewide, the number of farmers markets has more than doubled, with about 50 markets in operation. All of this has brought rural Idaho farm producers closer together with urban consumers.

"We like to buy local food because you know more of what you're getting in terms of how the food was grown and what kind of inputs were involved," says John Heimer, who was on the tour with his wife Marie. "Plus it

wormsetc.com

helps the local economy. Think globally and act locally, you know?"

Buying locally grown food at a farmers market can be more expensive than buying food in the grocery store, but Heimer says he's OK with that. "It does cost more, but you get what you pay for," he says. "You know more about what you're putting into your body."

"It's definitely a health issue," adds Sheryl Howe, who was on the tour with her husband, Gary Bahr, "knowing where your food comes from. Plus with fresh

Robert Hubner/WSU photo services

Left:
Student Arturo Ferrer Quintero finds *Pholiota squarrosa* in a wet North Idaho forest during a college foraging expedition.

Below:
Caps of *Amanita jacksonii* emerge from their cottony white mycelium. The reddish orange toxic fungus ranges from Canada to Mexico.

vegetables, the big issue is the flavor. They're significantly better. I get depressed when I go into the grocery store in January and February because nothing is fresh."

The Boise Co-op has been buying various kinds of mushrooms and heirloom tomatoes for a number of years, and the Boise community has a hearty appetite for those foods, says Airielle Jones, produce buyer for the Co-op. "There's a huge demand for morels and shitake mushrooms," Jones says. "And the porcinis did really well last year as well. The porcinis are cool because they're so big and meaty. I'm looking forward to more of those when Chris brings them in."

wondrous mycelia 63

Yancy/Deviant Art

"We also buy heirloom tomatoes—they just fly out the door," Jones continues. "With heirloom tomatoes, the taste is a big thing, and they have the original seed that hasn't been genetically altered. They've been grown for thousands and thousands of years in Europe. The flavor is there—they're a little weird-looking and juicy—but people know about heirlooms. They have so much more flavor than the tasteless tomatoes from Mexico you might buy in a grocery store. There's nothing like a fresh red tomato."

The nutritional value of wild mushrooms is another key selling point, Florence says. Morel mushrooms, for example, provide vitamin D and vitamin B, as well as a significant amount of potassium, calcium, magnesium and iron. Most other edible mushrooms have similar nutritional values. Specific edible mushrooms also have anti-virus elements to help with combating the common cold, for instance, and some even have anti-HIV therapeutic properties, Stamets says. "The key is to match the mushroom with the pathogen," he says.

If the current trends continue, the demand for natural foods will keep increasing, and Sweet Valley Family Farms will be well-positioned to handle the demand with its wild, natural foods collected by Chris Florence in the forests of Idaho and the Northwest. "I'm more than willing to pay more to support those smaller organic farms," Howe says. "In the long run,

Left:
The mystery and nutritional value of fungi lure foragers in the wild. Pictured: one serious mushroom hunter in Latvia, 2013.

Below:
Life's substrate web of microbiotic interdependence inspires 21st-century art. Pictured: Claire Burbridge's *Mycelium Universe 2*, 2014.

we have to look at the big picture and try to support them and keep them in business."

Florence thinks as the price of petroleum goes up, more farmers may look at changing to smaller, organic operations. Organic food does sell for a premium. "I think that's the way things are going to go because the current system we have is unsustainable," Florence says. "We're trying to make a living doing something we love. We're hoping to make a difference by providing healthy food for our communities."

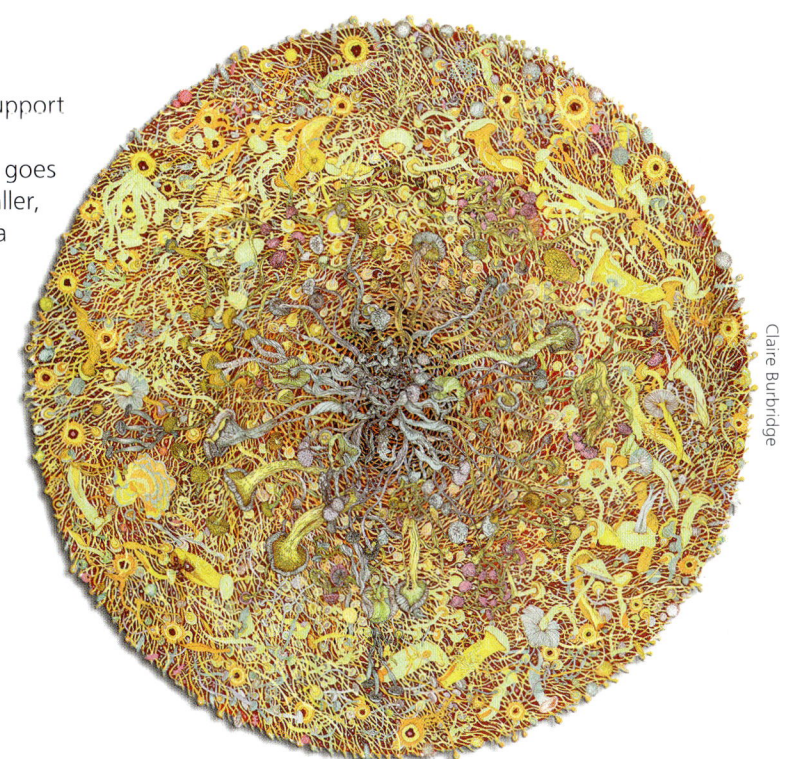

Vanishing of the Bees

You walk out to your backyard on a clear spring day and see your cherry tree is alive with movement and buzzing. Your brain rapidly calculates that your favorite climbing tree is host to a bazillion bees. The communication headquarters within your brain sends an immediate message to your legs ... RUN!!!

Bees are more than just the scourges of our childhood memories. Like morel mushrooms in the old-growth forests, bees are a keystone species—essential to agriculture, vital to keeping their greater ecosystem from catastrophic collapse. Without the industrious labors of these pollinating dynamos, flowing plants would struggle to exist.

Bees, domesticated and wild, pollinate close to 100 different food crops. Domesticated honeybees in the United States pollinate fruit, nuts, vegetables and other crops worth more than $15 billion annually. California's $4.8 billion almond industry relies on 1.4 million beehives. In Idaho, bee-dependent crops include onions, sugar beets, alfalfa hay and grapes.

But bees are disappearing and scientists struggle to understand why. In 1947, there were over 60 million managed U.S. beehives. Only 2.5 million survive. The mystery of "colony collapse" has been attributed to insecticides, herbicides and loss of natural habitat. Fungal microbes are among the suspects. So are bacteria, viruses and parasitic mites. One thing is for sure: our quality of life will be drastically altered if bees cease to work in our agricultural fields.

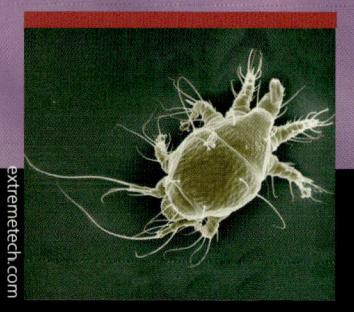

Nosema ceranae — extremetech.com

Beekeeper — Idaho Business Review

Mites on Bee Larvae — beespotter.org

Foulbrood Bacteria — Eshel Ben-Jacob

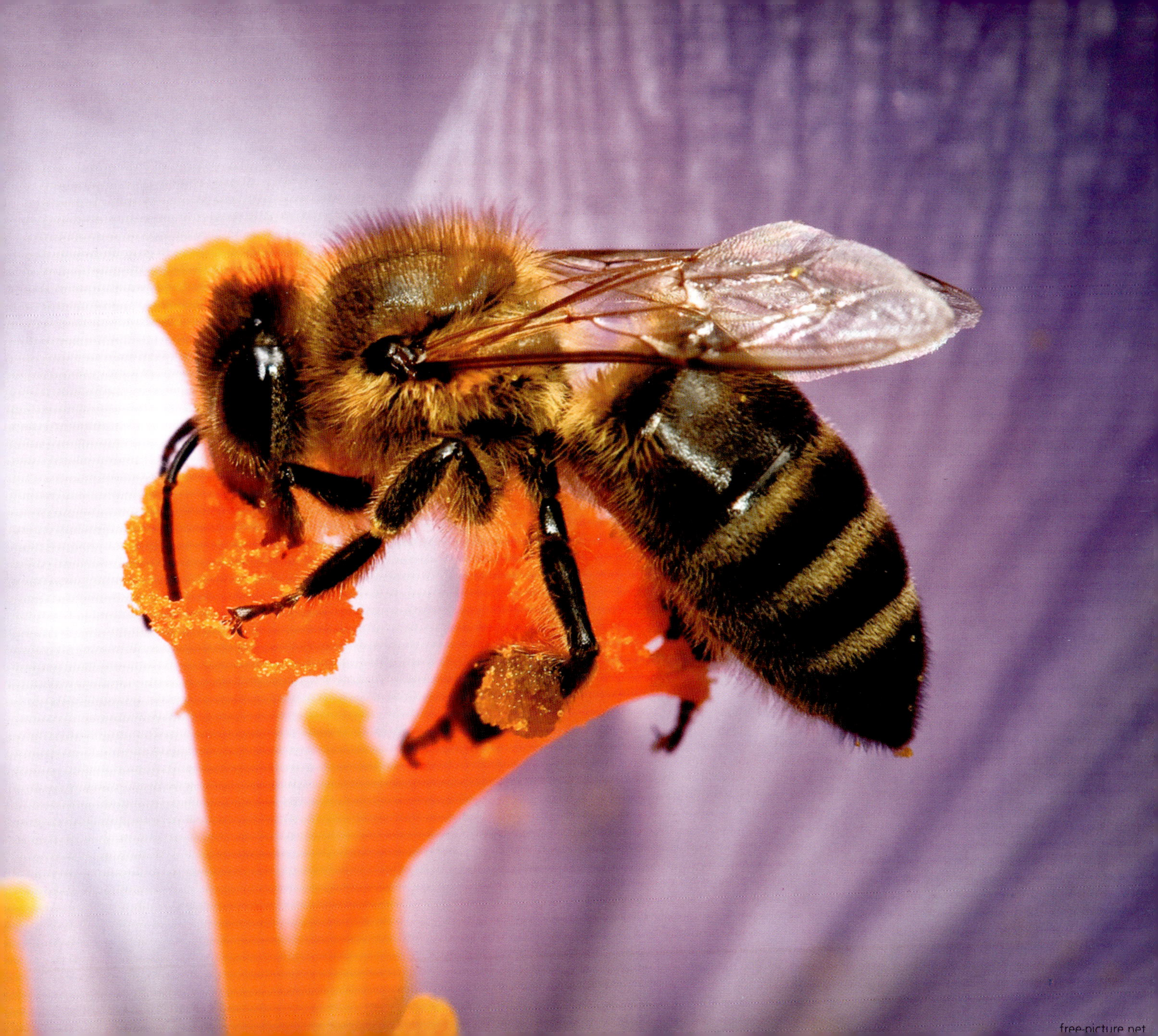

Langley Energy

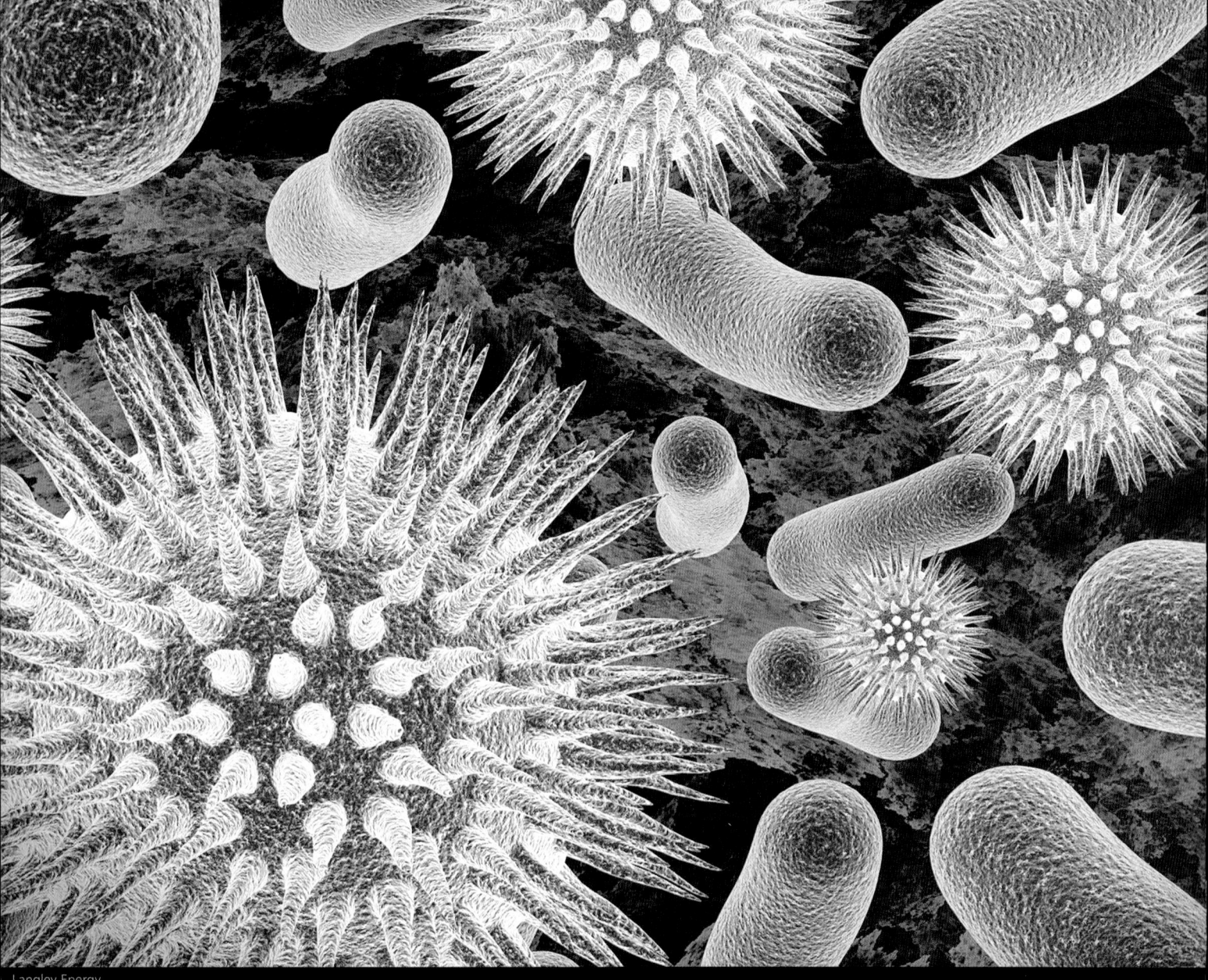

methane from microbes

converting cow manure into clean-burning fuel

3

Valtra Biogass

Left:
Methane-making microbes, called *methanogens*, thrive in the gaseous metabolism of dairy cattle. Below: Tin-plated two-handled milk cans predate the pulsating machines that make Idaho third in the nation in dairy.

Previous:
Methanogens resemble bacteria but differ in cell structure and genetic code. Microbiologists now group these methane-producers as archaea in the domain of Archaea, an ancient yet newly reclassified kingdom of life.

In the early 21st century in the Magic Valley region of southern Idaho, new dairy farms were popping up like daisies in the springtime. Many of the dairy farmers came to Idaho from California or the state of Washington, where they ran into rapidly escalating land values, high taxes and regulations. In the Magic Valley, which always has been an agriculturally rich region—the breadbasket of Idaho—dairy farmers were welcomed with open arms.

Word spread quickly that south-central Idaho could be the promised land for dairies, large-scale operations with thousands of cows. People could buy farms with water rights and begin operations in less than a year. Large dairies can bring odors, but the common adage from people accustomed to working in agriculture is "That's the smell of money."

One of the early transplants was Greg Ledbetter, who worked as a full-time veterinarian serving dairies in central and southern California before moving to Jerome, Idaho, in 1983. "When you can sell property down there for $100,000 an acre, at some point it makes sense to sell out and move your operation," Ledbetter told me in 2001 for a *High Country News* cover story about the in-migration of dairies.

The good news is that the Magic Valley has enjoyed rapid economic growth as the epicenter of Idaho's milk production. Over a 28-year period from 1980 to 2008, Idaho's dairy production increased more than 550 percent, propelling the state from the 18th largest dairy-producing state to the 3rd largest, behind California and Wisconsin.

The amount of milk produced annually in Idaho—approximately 13.2 million pounds—has an approximate value of $2.4 billion, according to the United Dairymen of Idaho. Dairy revenues surpassed potatoes as Idaho's No. 1 cash crop in 1997, and in a typical year, the dairy industry out-produces the revenue of potatoes by 2.5 times.

Danz Family

Below:

Archaea is ancient Greek for "ancient things." Archaea—a life form older than plants, older even than the Earth's continents—have been found in shale as ancient as 2.7 billion years.

Left:

Methanocaldococcus jannaschii is a heat-loving microbe found in hypothermal vents that bubble up from the ocean's floor. Widely used in high-temperature anaerobic digestion, the organism can tolerate heat up to 200 °F.

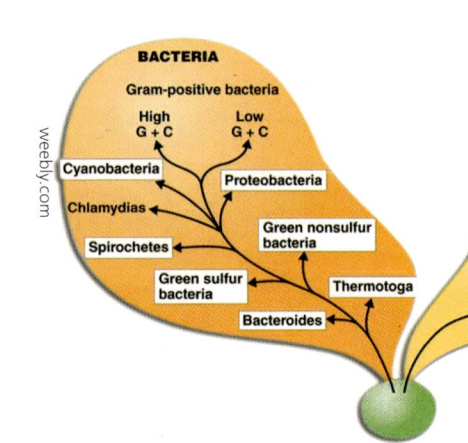

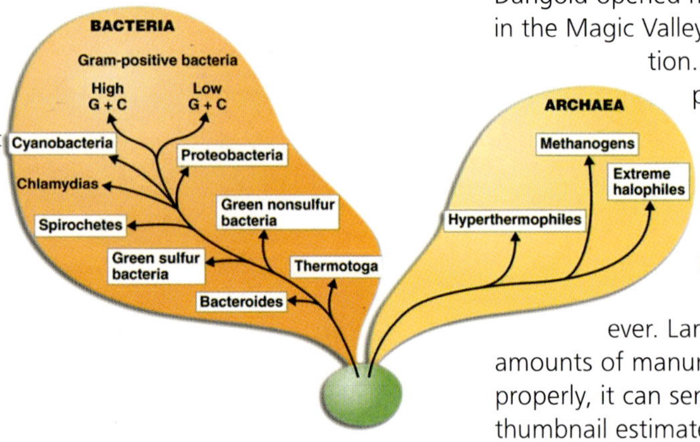

Adding value to the dairy expansion, Glanbia, Davisco, Kraft and Darigold opened multiple cheese, whey and condensed milk processing plants in the Magic Valley, bringing manufacturing facilities to the source of production. The milk-processing plants created more than 600 good-paying jobs that hadn't existed in the Magic Valley 15 years earlier. In 2012, New York–based Chobani Greek Yogurt announced plans to build a 1-million-square-foot, $450-million yogurt plant in Twin Falls, adding another 300 jobs to the mix. All told, the dairy industry in Idaho employs more than 13,000 people.

The rapid dairy expansion has had its drawbacks, however. Large dairies with thousands of milking cows generate large amounts of manure waste in a confined space. If that waste is not managed properly, it can send odors downwind and cause a nuisance to neighbors. A thumbnail estimate is that a dairy cow discharges 20 times as much waste per day as a human. That means a 1,000-cow dairy produces the same amount of waste as a city of 20,000 people with no sewage treatment. Decomposing liquid manure in waste lagoons can contain hydrogen sulfide, ammonia, volatile organic compounds and methane, a greenhouse gas of concern. State authorities do not require the waste lagoons to be lined with an impermeable material, creating concerns that manure waste could leak into groundwater and contaminate an important source of drinking water.

There are multiple ways to manage manure waste to reduce the environmental impacts, dairy officials say. Both liquid and solid manure waste can be used to fertilize crops. Manure waste also can be turned into compost. No matter how it's handled, manure waste can be a significant expense for a dairy operation. A dairy farmer with 8,000 cows could spend $1 million on waste management per year, officials say.

Dennis Kunkel Microscopy

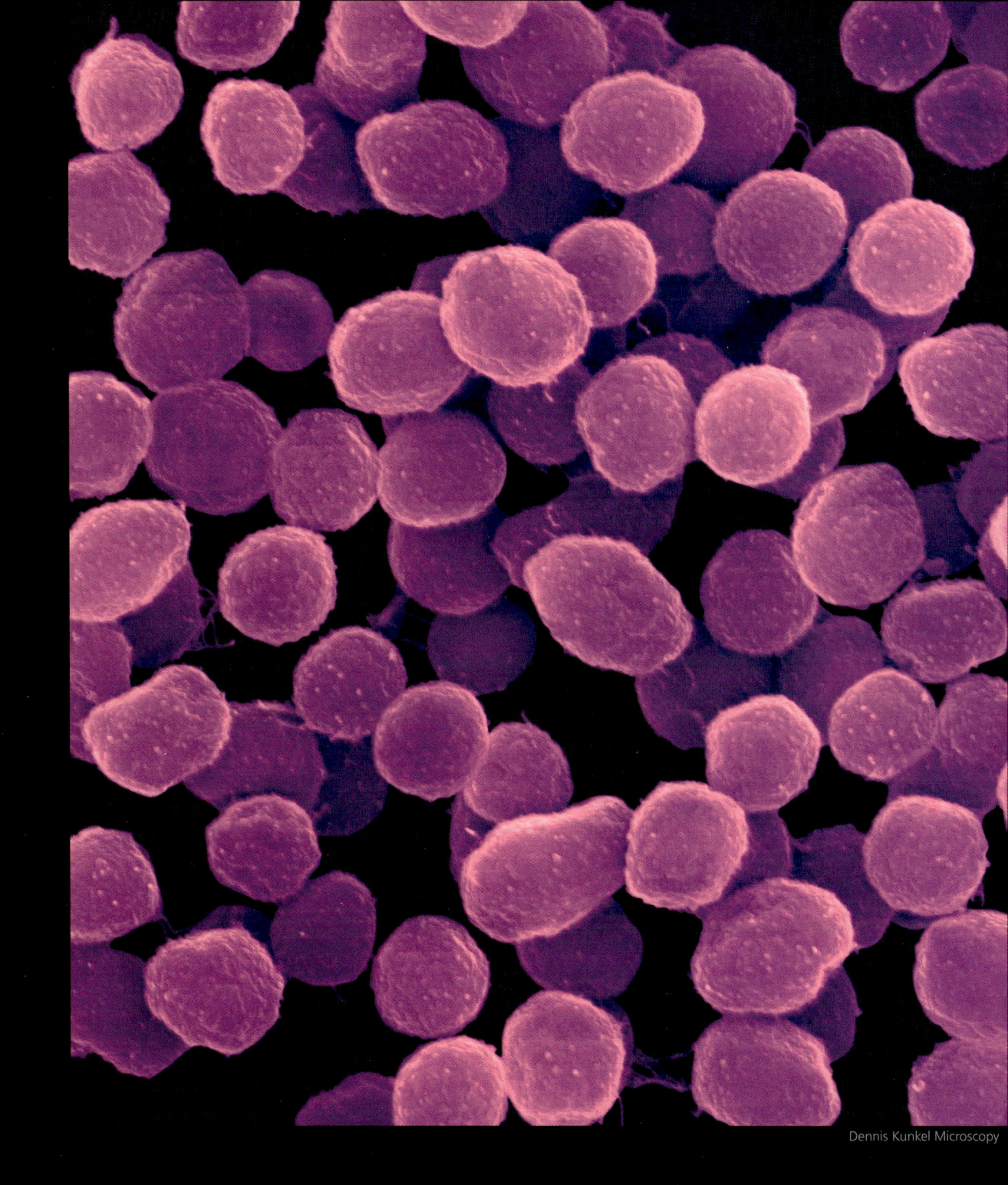

Below:
Methanogens are anaerobic, which means they don't need oxygen. They metabolize hydrogen and carbon dioxide, producing methane.

Right:
Anaerobic digesters decompose manure and convert the waste into fuel. One by-product is clean-burning methane. Another is a nutrient-rich liquid used for fertilizer.

University of Arkansas

"I see dairy waste as a great nutrient product—it's a great fertilizer," says Bob Naerebout, executive director of the Idaho Dairymen's Association. "I see multiple solutions rising out of dealing with manure waste. As the price of fertilizer goes up, people will see more value in manure as a natural fertilizer."

A multifaceted solution gaining ground in Idaho dairies is the process of converting manure waste into biogas or electricity through anaerobic digestion. It's an exciting, reemerging technology that not only converts waste to energy but also reduces dairy waste odors, cuts hydrogen sulfide emissions, and eliminates the threat of manure waste seeping into groundwater at dairies that send waste to anaerobic digesters.

"The sustainability of the dairy industry is going to depend on projects like this," says Jay Kesting, former director of renewable energy development for Western States Equipment. "There are 40,000 tons of manure processed here on an annual basis."

Western States spent $15 million to build the Rock Creek anaerobic digester project, the largest one built in the Magic Valley region so far. The project receives manure flow from three nearby dairies, with a combined total of 11,000 cows. The manure is piped on a continuous basis into the anaerobic digester

Organic Waste Systems

David Frazier

Left:
Dairy brings more revenue to Idaho than cattle or potatoes. Three counties of Idaho's Magic Valley—Jerome, Gooding and Twin Falls—lead the milk-cheese-yogurt bonanza with nearly 400,000 cows on about 360 dairies.

Below:
Methanosaeta are prodigious makers of methane. Species of this ancient genus typically inhabit the wetlands. As glaciers recede and tundra thaws into wetlands, the organism contributes to the greenhouse gases heating the Earth.

facility, where microorganisms work their magic in six 1-million-gallon tanks to convert the manure into biogas. At full production, the Rock Creek project produces 3.2 megawatts of electricity, enough to power 2,000 homes. "The power developed here is being used locally," Kesting notes.

Indeed, directly adjacent to the six digester tanks, all of which have a low-slung conical cover, a series of three-strand power lines carry electricity from the digester facility to Idaho Power's local grid running along a nearby highway. The whole digester facility is neatly tucked into the back corner of a dairy farm, about 10 miles southwest of Twin Falls.

Inside the quiet control room of a beige warehouse-like building, with a colorful graphic on his laptop Kesting shows how the anaerobic digester operation works. The computer software behind the graphic interface allows Kesting to control more than 180 valves, switches, pumps and other things at any time. On the left side are five reception tanks that store manure waste. The solids are separated by a sorting machine to produce a waste stream of 10 percent solids and 90 percent liquid.

A series of pipes carry the waste stream into five of the digester tanks (one of them is reserved for research). The computer screen shows how fast

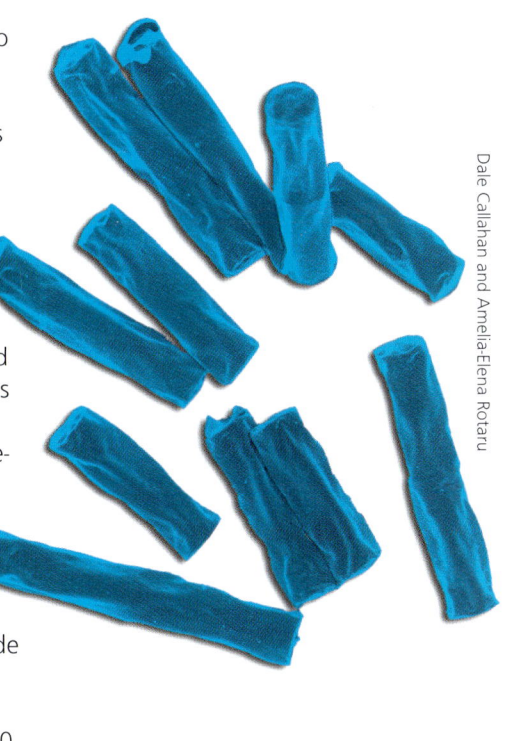

Dale Callahan and Amelia-Elena Rotaru

greenglancy.com

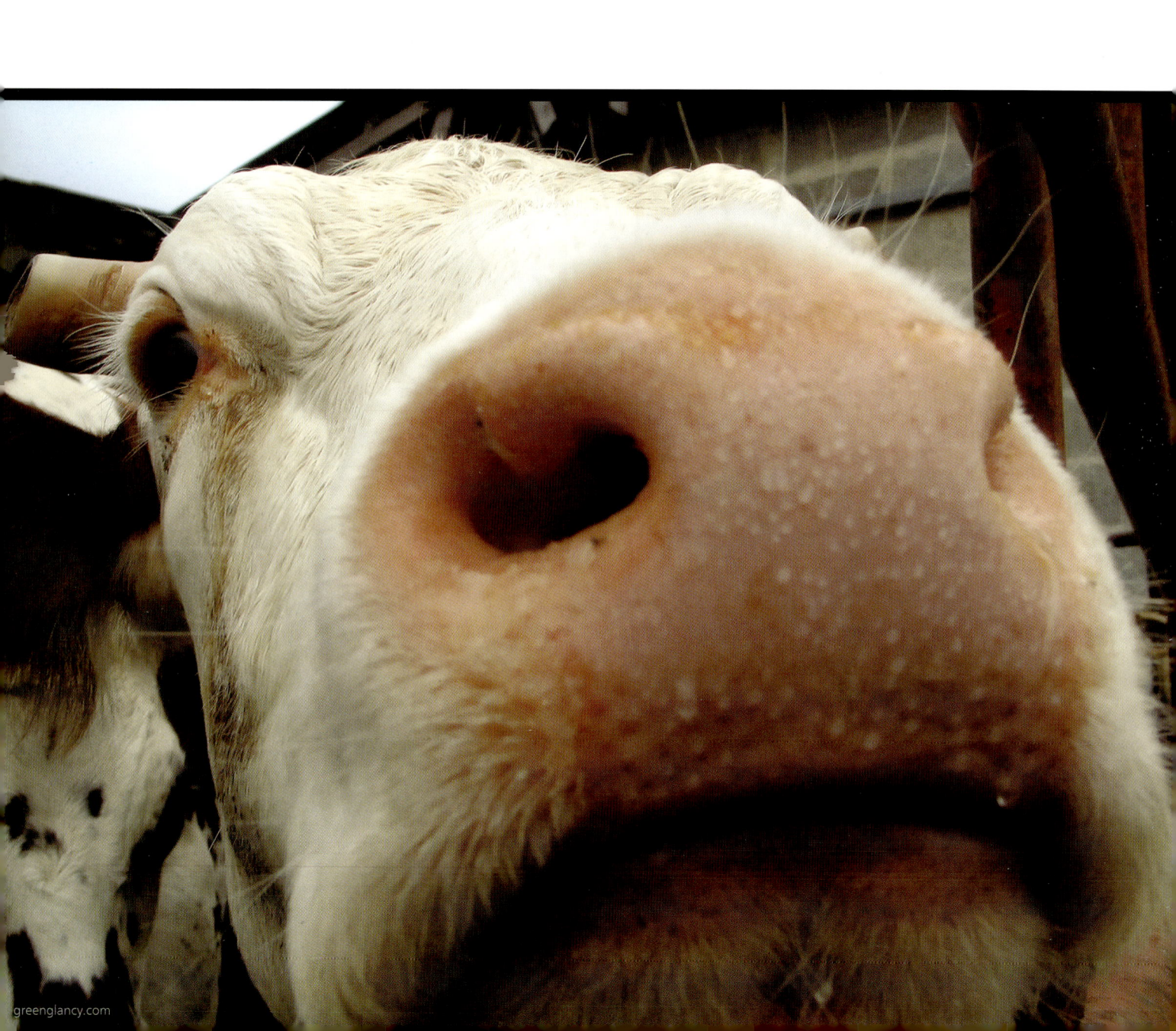

Left:
Cow flatulence? More methane passes through the front of the cow than the back. Through belching, mostly, cow methane accounts for about 24 percent of all the greenhouse gases emitted from livestock.

Below:
Methanogenesis is the final stage of four in the decomposition of organic matter. Inset: Filer's Rock Creek Dairy processes waste from 9,000 cows. Six 30-foot air-tight digester tanks each hold a million gallons.

the waste stream flows into the tanks in terms of gallons per minute, the volume of waste in each digester tank, the temperature of each digester tank (all of them are hovering in the 35.5-36.6 °C [96-98 °F] range) and the outflow of biogas to two large CAT 3520 engine generators.

Currently, the facility produces 11,600 kilowatt-hours (kWh) of electricity per day. That is enough electricity to power 305 average households based on the statistical annual consumption of 14,000 kWh, or 38 kWh per day. Kesting, who is dressed in a hoodie, a pair of jeans, and a green ball cap with the logo "Bison Engineering," has been up all night tinkering with the controls. The current problem is that the temperature in the digester tanks was raised too fast, which can cause an excess of foam. "It's a manageable situation," he says. Even so, running the digester is "complex," he notes. "You change one little thing, and it's like a butterfly effect."

Even though Kesting's title is renewable energy manager, he calls himself "the troubleshooter." He's the technical guy who knows how the facility works frontward and backward. During our interview, I can see his brain constantly working. He's not only thinking about the answers to my questions, but in the back of his mind, I can tell that he's consumed with bigger issues that kept him up all night.

methane from microbes 79

WHOI

Left:
So-called black smokers are the ocean's hydrothermal vents of an extreme kind of ecosystem. Deep-sea methanogens seek out the boiling heat. Bizarre creatures such as gutless eight-foot ice worms metabolize the methanogens.

Below:
The gut of a termite relies on methanogens to metabolize fiber. Termites pass an enormous tonnage of methane. The world's 2,000 species of termites emit an estimated 20 million tons of methane each year.

Generally, it takes 14 to 21 days for the methanogens to work their magic in the digester tanks before methane gas is produced off the top. The final output is 60 percent methane gas, 35 percent carbon dioxide, 2 to 3 percent oxygen, and trace amounts of hydrogen sulfide and nitrogen. With five tanks operating at once, and the manure slurry always bringing in more waste into the digester tanks, it's a continuous process of converting waste to energy.

Kesting heats up the temperature of the manure slurry by 11 degrees Celsius (20 °F) before it's pumped into the digester tanks to create the best environment for the methanogens. The term *anaerobic digestion* means there is no oxygen present in the sealed tanks. The methanogens thrive at higher temperatures. Charlton likens methanogens to the temperament of a 2-year-old. "They're very fragile and they love their sugar," he says. "The more you give them, the faster they go. If you give them a loving environment and a home, they'll do everything else on their own."

In truth, Kesting says, the digester tanks are not completely devoid of oxygen. He injects a tiny amount of carbon dioxide into the digester tanks to keep nitrogen and hydrogen sulfide emissions at a manageable level. After the biogas is sucked off the top, there are two by-products—liquid waste and solid waste. The liquids can be applied to farm fields as fertilizer, and the solids are separated for use as peat moss or animal bedding material. Kesting shows me a bin of the solids and I grab a handful. The material is soft, pliable and warm, with no residue.

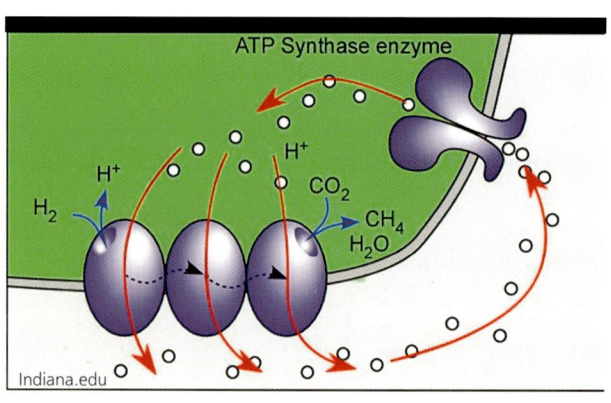

From time to time, Charlton draws a sample of the methanogens at the Rock Creek facility to see which species are present. "When you look at them under the microscope, they look like little blobs and natural colonies," he says. "I look at the feed stock, the shape of the organisms, the colonies and figure out what population it is. There are about 40 different species of methanogens that function in Jay's digester, and I've been able to positively identify six or seven of them so far."

Methane-forming microbes are some of the oldest life forms on Earth. Since 1977, scientists have grouped them in the kingdom or domain of Archaea, meaning ancient. Methane-forming archaea flourish in heat. They're "predominantly terrestrial and aquatic organisms and are found naturally in decaying organic matter, deep-sea volcanic vents, deep sediment, geothermal springs, and the black mud of lakes and swamps," explains author Michael H. Gerardi in the book *The Microbiology of Anaerobic Digesters*.

Left:
Bacteria in the gut of a cow make hydrogen and carbon dioxide. Methanogens convert these molecules to methane. Bottom: Duke University researchers mix food waste and sewage in formulas for anaerobic digestion.

Below:
Carbon traces in reefs and fragments of limestone indicate that archaea may have appeared a billion years after the Earth's creation when conditions were hostile to other forms of life.

Methane-forming archaea also are found in the digestive tracts of humans and animals, especially in the rumen of herbivores such as cows, goats, sheep and deer. Two species of the methane-forming archaea, *Methanobacterium formicium* and *Methanobrevibacter arboriphilus*, are the primary species found in anaerobic digesters, Gerardi writes. These species help degrade organic compounds, and they produce methane while doing so. Methane from cow manure has been mixed with carbon dioxide and hydrogen sulfide to make powerful fuels.

From a business and economic point of view, a key aspect of the Rock Creek anaerobic digester project is that Western States Equipment officials were able to negotiate a favorable power rate with Idaho Power Company, the state's largest utility. Western States signed a 15-year contract with Idaho Power to receive 6 cents per kilowatt-hour for the electricity the digester produces, Kesting says. The power-reimbursement rate can make a huge difference between whether a digester project is a profitable or a losing business proposition.

"We get a 2-cent differential in the rate if we make more power during the day than at night," he said. Idaho Power prefers to receive the most power when customers need it, and Kesting runs the plant to capture that premium.

Working on the business and economic side of the anaerobic digester projects is critical to ensure that more projects can be built in the future, Kesting says. "If we can't make money on it, businesses are not going to invest in it. It's not like we're building the freeway system here or putting a man on the moon. There's no government subsidy for these types of projects."

Naerebout of the United Dairymen of Idaho agrees. "The biggest issue with digester projects is the economic side of it," he says. "So far,

sunyorange.edu

methane from microbes 85

we've had a few large conglomerates coming in here and building digester projects. We'd really hit a home run if our dairy producers could invest in anaerobic digester projects and build their own."

At some point in the future, it's likely that either the private sector or the U.S. government will get more involved in trading carbon credits, as they do in Western Europe, California and Canada. Companies that produce greenhouse gas emissions with coal plants, for example, can purchase carbon credits by investing in renewable energy projects. That could make renewable energy projects, and anaerobic digester projects in particular, more valuable.

A new scientific breakthrough involving plastics could add even more economic value to anaerobic digester projects. Erik Coats, an associate professor of civil engineering at the University of Idaho, has been working with his graduate students to develop biodegradable plastic products from manure. Coats and his team of researchers have created small-scale digester projects in a U of I lab, using manure from the U of I dairy. They've figured out how to use naturally occurring archaea in manure to produce a plastic called *polyhydroxyalkanoate* (PHA).

Coats explains that PHA is similar to polyethylene and polypropylene. "But in contrast to these petro-plastics, PHA is biodegradable and can be produced from otherwise unwanted organic waste," the professor wrote in a business column in the *Idaho Statesman*. The research could lead to broad-scale production of biodegradable plastics at a facility like the Rock Creek digester. That would lead to another significant income stream.

If Coats' new research project proves to be viable at a large scale, he hopes to sell the technology to a private sector partner for broader development. The project is so promising that the United Dairymen of Idaho, Idaho

Left:
Microbes of many kinds—archaea, bacteria, protozoa and viruses—work in the biochemical symbiosis that allows herbivores to metabolize grass. Pictured: Red cell walls join a cluster of cow-gut microbes.

Below:
Calendar art from Tanzania pictures the top of an underground biofuel tank for the recycling of cow manure, 2008. A preindustrial technology, anaerobic digesting dates back to the 17th-century work of Irish chemist Robert Boyle.

National Laboratory, National Science Foundation and Idaho State Board of Education are all supporting it. "It's pretty exciting stuff," Coats tells me as we tour his lab at the U of I. "The plastics side of this could put a whole new economic slant on things."

Kesting is hoping that there will be potential for another 20 to 25 anaerobic digester projects in southern Idaho, between the Magic Valley area and dairies in the Treasure Valley's Canyon County. A development company could look at building more anaerobic digester projects in states with much higher electric power rates, such as New York, Wisconsin and California, where the economic feasibility would be higher. "We're taking things in the right direction," he says.

Perhaps best of all, the neighbors downwind of the Rock Creek dairy are breathing much easier than they did when the dairy rush was occurring more than 10 years ago. This is in part because the Rock Creek anaerobic digester project does its work in an enclosed facility that reduces odors significantly. Ed Smith, a retiree who lives with his wife Phuong in Cedar Draw, south of Twin Falls, says nearby dairies have changed. Improved management practices, including the Rock Creek digester, have reduced odors. "It's not an everyday issue like it was before," he says. "We can take walks down the lane now in the evening without encountering a wall of odor. In fact, now, we don't even think about it."

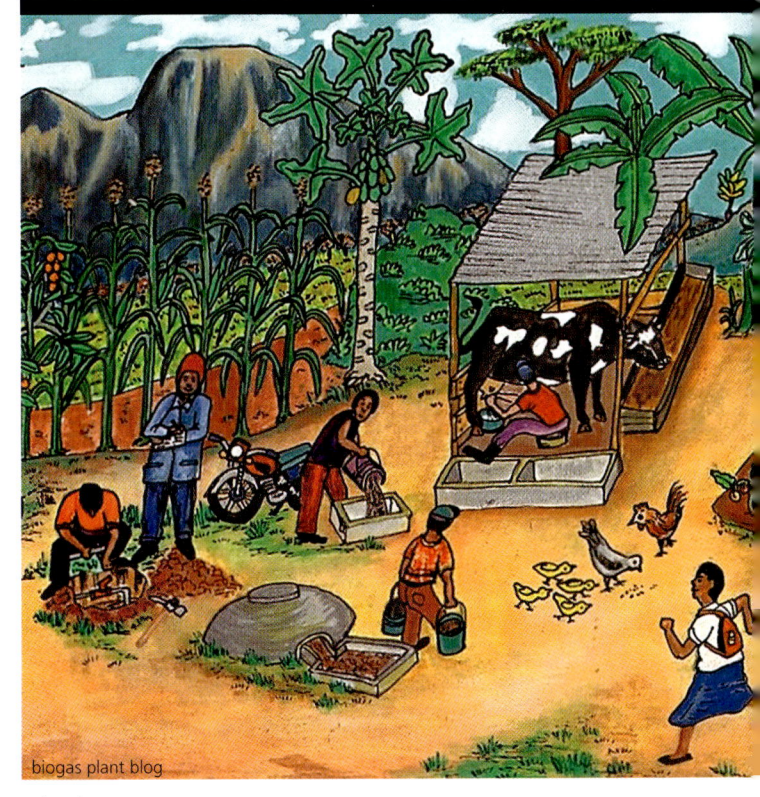

Thriving in the Extremes

Extremophiles are organisms that adapt to the extremes. Some flourish in vents that boil the ocean with plumes of hydrogen sulfide. Some cope with radiation. Some eat petroleum. Some metabolize heavy metals below the Arctic's ice.

Extremophiles are a fascination for science. Once classified as bacteria, extremophiles are now mostly grouped among the archaea from an older branch of the tree of life.

Idaho's sulfur hot springs swirl with extremophiles. At Lidy Hot Springs in Idaho's Beaverhead Mountains, where temperatures top 54 °C (130 °F), methanogens populate pools that may resemble underground water on Mars. Extremophiles of many kinds also color with vibrant hues the sulfur pools of Yellowstone National Park.

Pictured: NASA's Kevin Hand hunts for microbial life beneath a frozen lake in Alaska. Insets from left: small worms feed on bacteria in the deep Atlantic; methane-eating ice worm from the Gulf of Mexico; marine bacteria that metabolize sulfur; subterranean microbe from Antarctica's Lake Whillans.

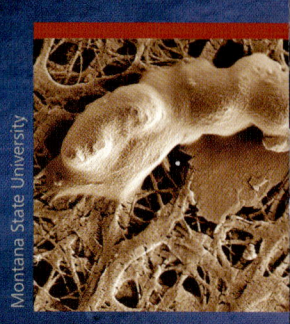

Mark Thiessen/National Geographic

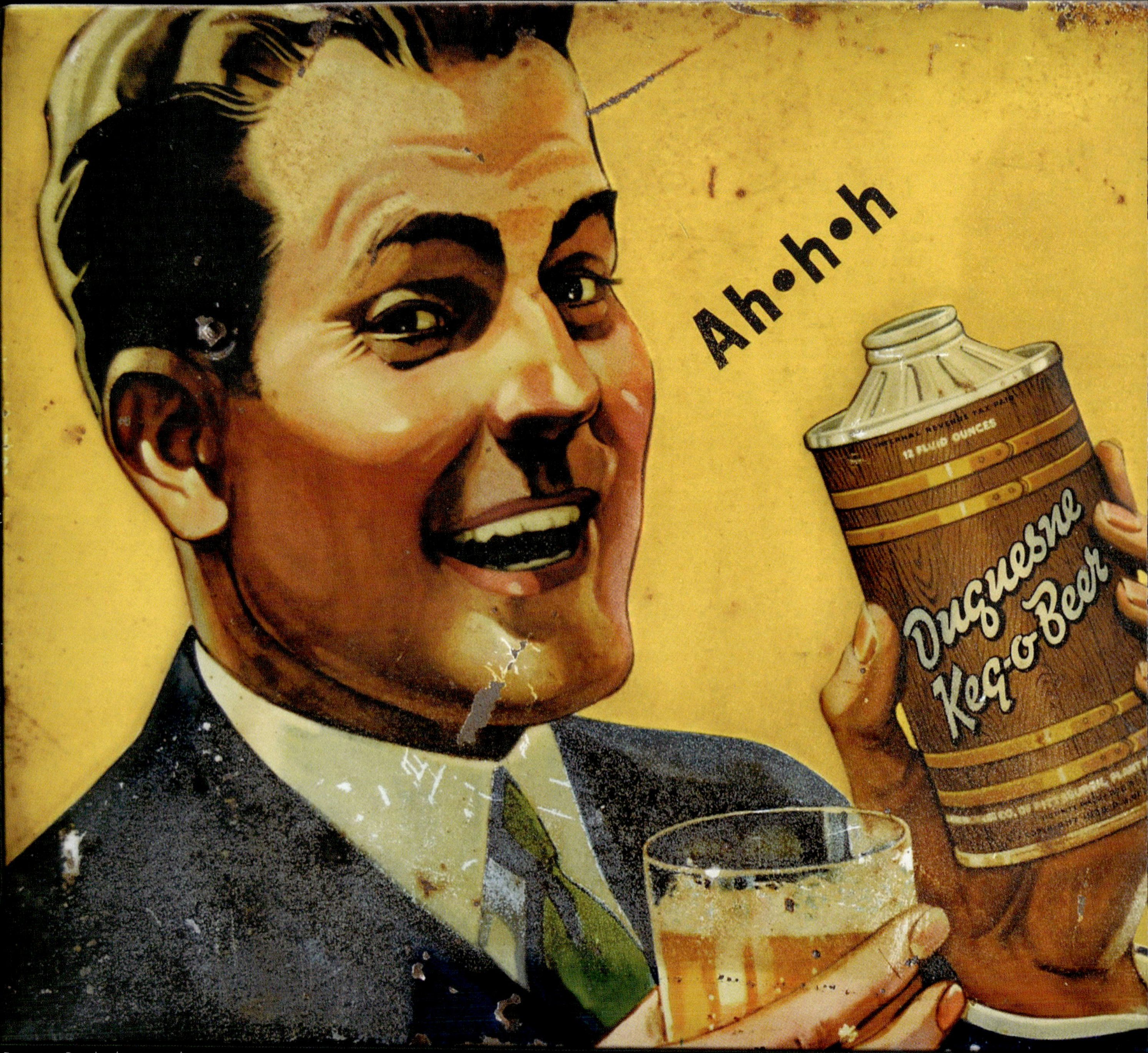

Duquense Brewing/garageartsings.com

the magic of yeast

and the rise of craft beer

4

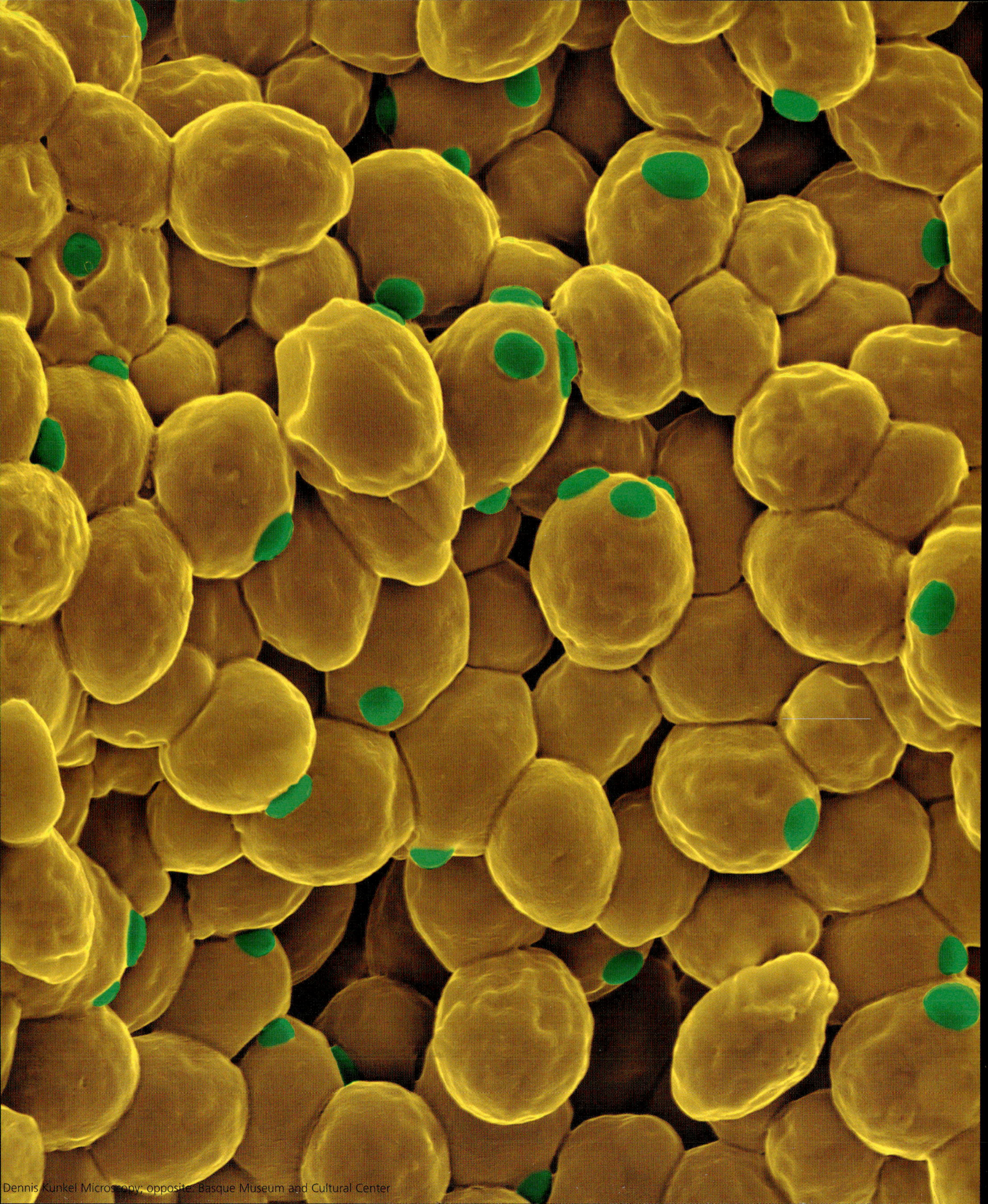

Dennis Kunkel Microscopy; opposite: Basque Museum and Cultural Center

Left:
Baker's yeast and brewer's yeast are common names for ancient strains of the fungus *Saccharomyces cerevisiae*. Yeast, cultivated by humans for 10,000 years, may be among the first organisms successfully domesticized.

Below:
Beer bottle cap recovered from under the floorboards of a pioneer house on Boise's Grove Street. Previous: Duquesne Brewing of Pittsburgh, founded in 1899, marketed pilsners in Boise with advertisements on rail cars.

Matt Ganz of McCall, Idaho, has a brewer's fascination for yeast and a hearty thirst for high adventure. A native of Spokane, Washington, the bearded Ganz went to free-thinking Evergreen State College in Olympia. After finishing college, he immersed himself in the tough, gritty world of wildland firefighting. Wearing the trademark Nomex yellow shirt made of flame-resistant material, red bandanna, green pants and a helmet, he dug line and fought fire with the Midnight Sun Hotshots, stationed in Fairbanks, Alaska, for five seasons. Then he ratcheted it up a notch to become a smokejumper, those gutsy people who parachute from low-flying airplanes into new fire starts in wild, rugged mountains in the West and try to nip the blazes in the bud.

Imagine the excitement and adrenaline of standing on the edge of the open door in a smokejumper plane and hearing the spotter say, "Get ready!"

Ganz did that 33 times on real fires for the U.S. Forest Service in McCall, and he notched 53 practice jumps over five years. Besides the excitement, "I felt a deep sense of fear, but I didn't that let that stop me," he says. "I would acknowledge that fear and funnel that energy into the task before me."

"I loved that job," Ganz goes on. "It was really fun and exciting. Smokejumping is definitely considered to be the top tier for western wildland firefighting. I reached a pinnacle that only a few other people have achieved."

Nowadays, Ganz faces a different kind of challenge, but one that is equally daunting and electrifying—building a new business from scratch, learning how to make quality beer and carving out a niche in the craft beer movement in southwest Idaho. Ganz and his wife, Ellen, and business

magic of yeast 93

partners, Matt and Jennifer Hurlbutt, took the leap in 2009 by opening Salmon River Brewery in McCall.

Ganz says the adrenaline rush (and, quite candidly, the fear) associated with taking a business risk, plus learning how to make quality, craft beer was similar to the smokejumping experience. "A lot of jumpers wonder, if I quit, how can I ever find something just as exciting?" he says. "I feel that I've found something that I'm just as passionate about as smokejumping. The basic work of making beer, starting a business from scratch and seeing the community embrace it took a ton of passion and energy. To see it succeed is extremely exciting. I've felt that fear, but I'm not letting it stop me."

The craft beers offered by Salmon River Brewery—Udaho Gold, PFD Pale Ale, Shiver IPA and Buzz Buzz Coffee Porter, among others—took McCall by storm. The small mountain community already had another brew pub, McCall Brew-

Topeka and Shawnee Public Library

Library of Congress

94

Left:
Hatchets became symbols of prohibition after crusader Carrie Nation, in 1901, smashed kegs in Kansas saloons. Bottom: Protesters march for legalization at a New York "beer for taxation" parade, 1932.

Below:
Salmon River's Matt Ganz loves the earthy aroma of hops. The McCall brewmaster adds Idaho hops to his tanks at the last stage of fermentation—a method called dry hopping.

ing Company, but something about Salmon River Brewery's approach (maybe it was the inclusion of a sushi bar?) made it hugely popular from Day One.

"It's definitely one of the coolest things that's happened in McCall," says David Bingaman, a river guide and director of the Payette Avalanche Center in McCall. "When they opened, everyone was really excited about it. The Hurlbutts and the Ganzes were really well-known in McCall. They're like 'your bros' so everyone wanted to support them. Plus, it had such a different feeling than anything else in McCall. You walk in there, and you're like, this is such a cool place."

Later, Bingaman would play a support role in the brewery, taking things to the next level. More about that in a moment.

The rise of the craft beer movement began in the late 1970s in California and gained steam in the mid-1980s in Seattle, Portland, Boston, and Boulder, Colorado. In the 1990s, several microbreweries emerged in the Treasure Valley, including the Ram, Sockeye Brewing, Table Rock Brew Pub and Highlands Hollow. Now, in the second decade of the 21st century, it is rocketing forward with double-digit percentage growth practically everywhere in the United States. In the Boise area, many new craft breweries have opened in the past few years, including Payette Brewing, Crooked Fence, Kilted Dragon, Slanted Rock, Ten Barrel, PostModern Brewers and Haff Brewing. There's even a website promoting the Boise Ale Trail, from

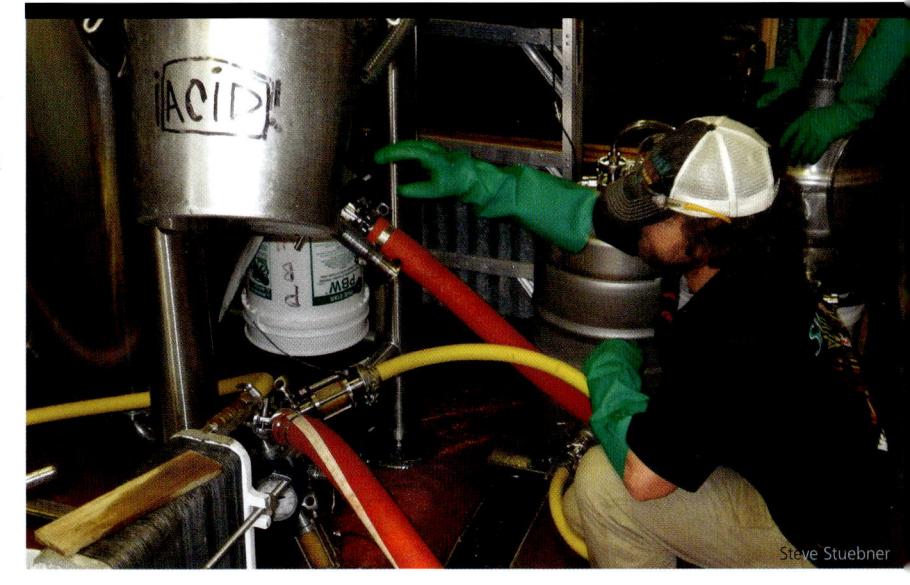

Dennis Kunkel Microscopy

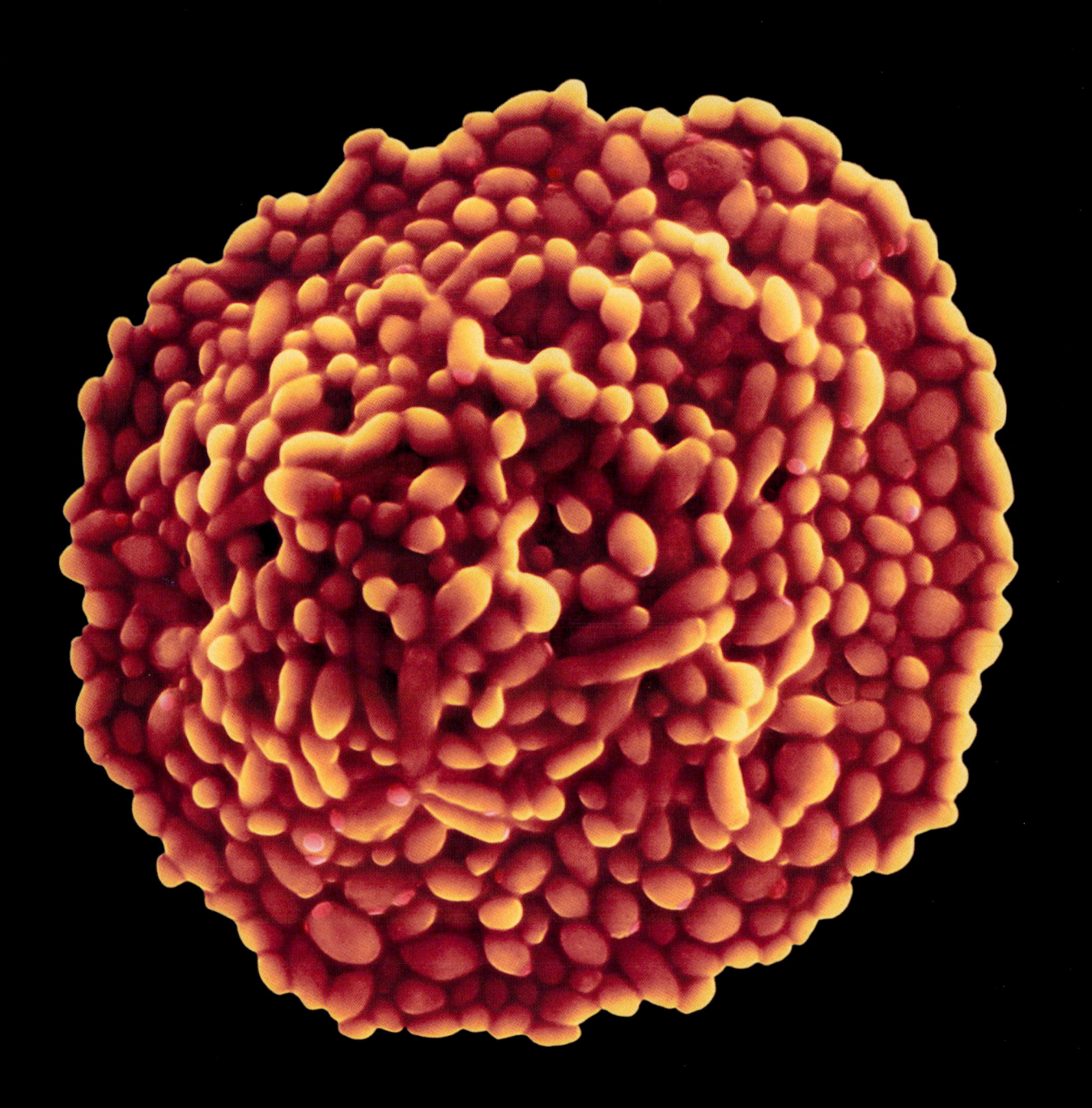

Left:
Scientists have classified more than 1,500 species and strains of yeast in the kingdom of Fungi. *Saccharomyces cerevisiae*, pictured, is the species most pervasive in the making of bread and beer.

Below:
Two different strains of yeast make beer into lager or ale. For lager, the yeast brews colder and more slowly at the bottom of the fermentation tank. Pictured: Magic Yeast of Chicago, about 1900.

the Ram in east Boise to Slanted Rock in Meridian, showing how all of the small breweries are working together to share the newfound enthusiasm for quality beer.

"From a craft brewing perspective, the American public has demanded that they want better beer," says Jeremy Canning, co-owner of Kilted Dragon. "That's part of what drives the hunger for more beer choice. Compared to Oregon and Washington, Boise is really just starting to get on the map."

Boise is primarily "an IPA town in my humble opinion," Canning says, but the Kilted Dragon also pushes a Knuckle Dragger Porter that's a mix of chocolate-roasted malts, brown sugar and fresh banana flavor. "This porter is equal to a heavy metal band taking the stage in tuxedos—loud and in your face, yet smooth and refined with a hard-to-describe sense of class," he says. "If I can get it in someone's hands, they're going to like it."

The craft beer craze also is happening on an individual home-brewer scale. No one knows how many home-brewers exist in Idaho, but according to the American Homebrewers Association, there are at least 1 million Americans brewing beer at home. That's considered to be a conservative estimate. Clearly, people are seizing on the opportunity to make their own beer, joining the quest to invent something special and unique, and share tasty suds with their friends.

"The home-brewers have a lot to do with the renaissance," says Patrick Orr, a former beer columnist and reporter for the *Idaho Statesman*. "People realize they can make a tastier beer in their own home, and some of them decided to take the next step and build a business around their beer."

That the "Big Three" beer manufacturers in the United States—Anheuser-Busch, Miller and Coors—haven't responded to Americans' demand for better-tasting beer leaves an opening for entrepreneurs, Orr says. "Craft beers

countryconsultant.com

just taste way better. You go to Belgium, Germany, Ireland, and drink their flavorful beer, and you wonder why that kind of quality beer isn't available in the U.S. on a broader scale. But now, with the craft beer movement, we can buy some good quality beer at the grocery store, and we can do that in Boise."

In the documentary film, *Beer Wars*, Director Anat Baron explains how the craft beer movement is a David vs. Goliath battle for the beer consumer. A lot of Americans love Bud Light and Coors Light. They dominate the sales. But those brands taste as if they've been watered down, according to critics. Jim Koch, founder of the Boston Beer Company, maker of Samuel Adams beer, says he set out "to make the best beer in America." Beer-makers needed to set higher standards, he said.

"Unfortunately, all of our beer was coming from Anheuser-Busch, Miller and Coors. It's as if all we knew about food came from McDonald's," Koch says in *Beer Wars*.

Yeast, a well-known single-celled microorganism used to make bread and the like, lies at the very core of the craft beer movement. It also lies at the heart of making beer, which dates back to at least 3000 BC in Mesopotamia. Early beer-makers had to dry malts over a wood fire, so apparently many of the early beers tasted quite smoky. No one understood the

National Museum of Medicine

Left:
Louis Pasteur's beer vat, about 1880. Air-tight fermentation, Pasteur discovered, increased the sugar content and kept bacteria from spoiling the brew.

Below:
Craft beer is big in Boise, where adult night at the Discovery Center featured the science of brewing. Bottom: Philistine beer jug, about 800 BC.

intricacies of the fermentation process in those days, so a beer batch would spoil quickly—in a matter of hours or days.

Louis Pasteur, one of the three founders of microbiology, is credited with discovering, in the 1850s, that yeast was a single-celled living organism. He proved in 1857 that yeast was the causative agent in the fermentation process and the making of beer. He was working on curing the "disease of beer" at the time. This was a hugely important discovery, according to beer connoisseurs.

"It is not exaggerating to suggest Pasteur made the greatest advances of anyone in the history of beer, and that these breakthroughs and others led to some important advances for the whole of civilization," wrote Chris White and Jamil Zainasheff in the book, *Yeast: The Practical Guide to Beer Fermentation*. "His studies into beer and wine fermentation paved the way for his later work on anthrax, rabies, cholera and other afflictions, which led to the development of the first vaccines."

For centuries before Pasteur's discoveries, the early beers fermented naturally in open containers by yeasts that were carried by airborne dust. In other words, the fermentation happened largely by accident. Pasteur's discoveries indicated the importance of controlling the types of yeasts and microorganisms in the fermentation tank.

"Pasteur traveled from brewery to brewery in the late 1800s and offered his services to inspect their yeast cultures, and gave the breweries a passing or failing grade," according to White and Zainasheff. He traveled throughout Europe doing this. When Pasteur explained the importance of yeast to English brewers in the late 1800s, they were inspired to hire chemists as senior-level staff members. "These brewing

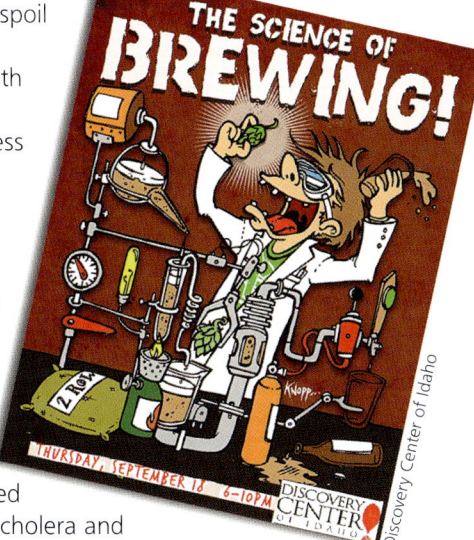

Hanay

magic of yeast 99

Science Source

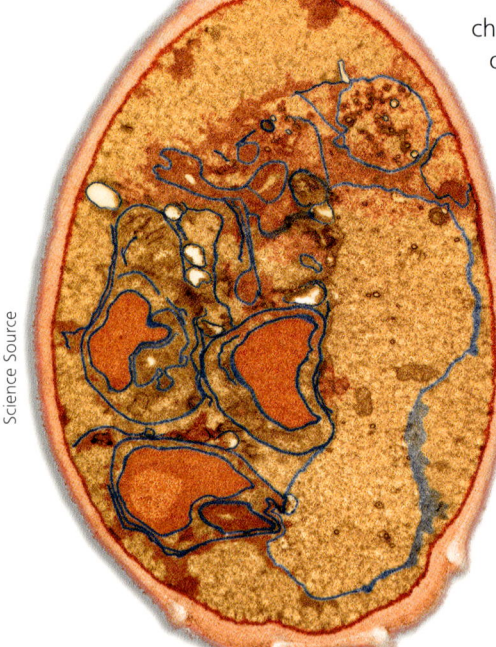

chemists became highly sought after and also became the highest-paid members of the brewery staffs," the authors said.

Brewers quickly began working on isolating their favorite yeast strains for making beer. In 1883, the Carlsberg Laboratory in Copenhagen isolated the first lager yeast strain. Its scientific name was *Saccharomyces carlsbergensis* or *Saccharomyces uvarum*. That caused lager beers to take off in popularity, because most ale fermentations still contained some "wild" yeast (meaning it was exposed to outside air and could be contaminated with a variety of bacteria and yeasts). "The resulting beer, even if it was acceptable at first, had a short shelf life before it went bad," the authors wrote. "Lager beer had a longer shelf life, which meant a larger distribution area and increased sales."

Recently, I toured the Payette Brewing Company in Garden City to learn how beer is made. Mike Francis, owner of Payette Brewing, explained the process as he gave a tour of the bright shiny back room of his brewhouse to a group from the City Club of Boise. Payette has a large space for making beer, with brew kettles and large stainless steel cone-shaped fermenters, and an open, wood-paneled tasting room in the front of the business.

Beer has four main ingredients, Francis explains, water, yeast, malt and hops. "Yeast plays the biggest role in the flavor component," he says.

A former industrial engineer for Boeing, Francis went to brewing school for three months to learn all about making beer. It's been trial and error ever since. Payette Brewing purchases a whole suite of different malts to make their different beers, about 20 of them in all. They have pale malts for the pale ales and chocolate malts for the stouts. They crush the grains and put them in a mash tub with water. Then, in a brew kettle, the malt is heated to a boil for about an hour or so. This is what's called the "hot side" of the beer-making process. "You want to eliminate all of the bacteria," Francis says.

Left:
Yeast is a single-celled fungus that metabolizes sugar, yielding alcohol and carbon dioxide. Beer, wine and whiskey all begin with yeast fermentation.

Below :
Pridefully local Payette Brewing of Boise strives to "payette forward" by hosting events for nonprofits and donating proceeds from its Busted Wagon holiday beer.

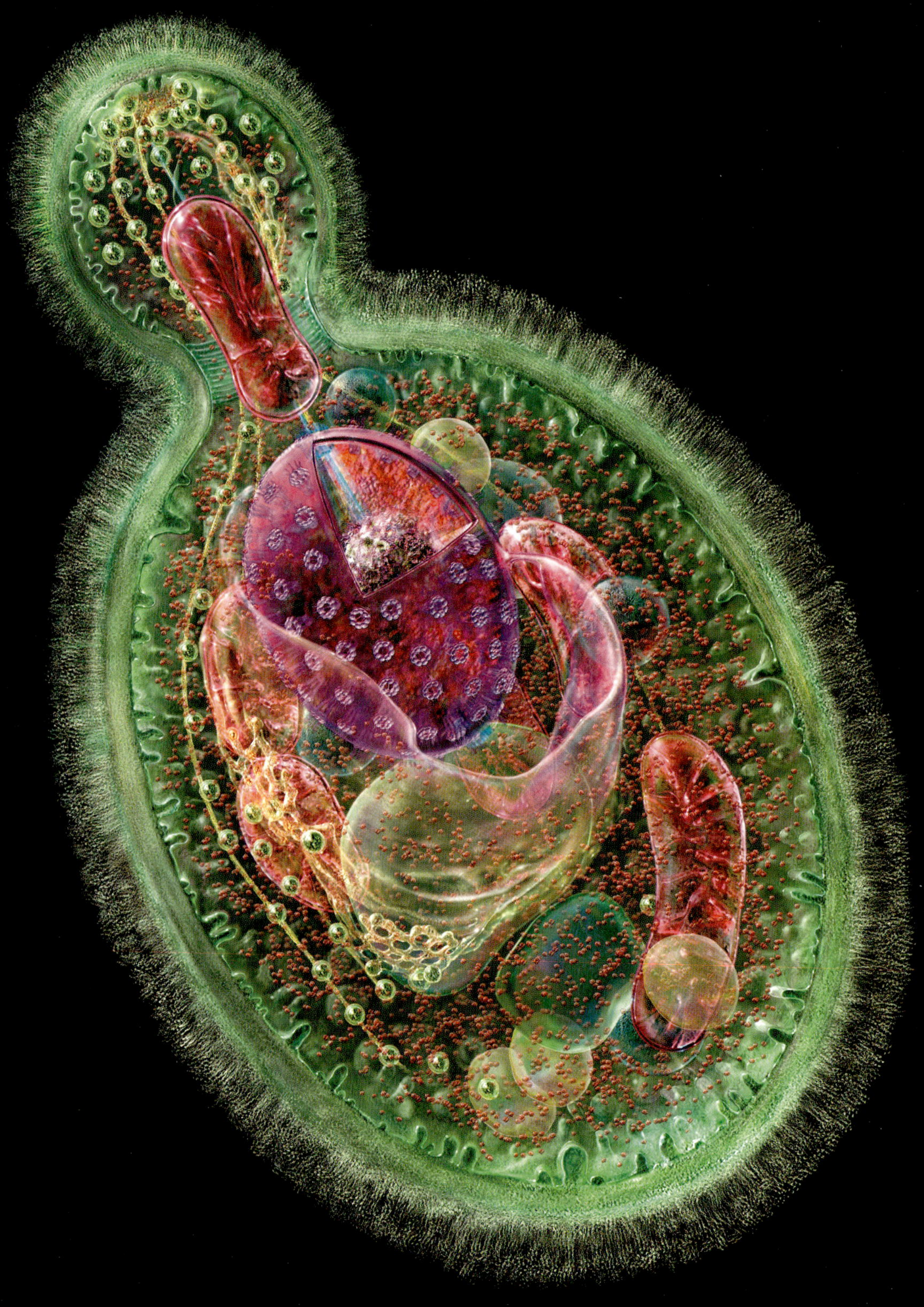

Science Source

Left:
Brewer's yeast reproduces by asexually "budding." The nucleus splits from parent and migrates into the "bud" of the daughter cell. The bud continues to grow until it detaches by forming a separate cell.

Below:
Yeast fortified with vitamin D inspired Schlitz to market its brew as a health food in 1936. "Beer is good for you," said a magazine advertisement, "but Schlitz, with Sunshine Vitamin D, is extra good for you."

While the kettle boils, he adds hops to the mix. He adds more hops for a pale ale and less for a brown ale. All of the beer recipes have precise amounts of each ingredient so they all turn out the same. After the malt and hops have been boiled long enough, the liquid known as "the wort" is transferred to a fermenter, where Francis adds liquid yeast to make beer. This is known as the "cold side" of the process. Each beer's recipe calls for a specific amount of yeast to be added and an ideal temperature for the fermentation process; for ales, it's usually between 65 and 72 degrees Fahrenheit. For lagers, it's in the range of 40 to 55 degrees Fahrenheit.

At most breweries, Payette Brewing included, the yeast interacts with the wort in an oxygen-free setting inside a sealed fermentation tank. This allows the brewmaster to completely control the brewing process. Typically, a brewer may add a smidge of oxygen to the wort, prior to going into the fermentation, to speed along the interaction of the yeast with the sugars of the malt.

The wort contains healthy amounts of nitrogen, sugars, minerals and vitamins, and the yeast cells consume all of those things during the fermentation process to make alcohol. In fact, the yeast cells survive in the oxygen-free environment by making alcohol or ethanol and carbon dioxide. That's the magic.

"Upon inoculation into the wort, the cells first utilize their glycogen reserves and any available oxygen to revitalize their cell membranes for optimal permeability and transfer of nutrients and sugars," White and Zainasheff wrote. "The cells rapidly absorb oxygen and then begin to pick up sugar and nutrients from the wort. Some of these compounds easily diffuse across the cell membrane and some require yeast transport mechanisms. Because yeast utilize some sugars more easily than others, they take up sugar in a specific order, with simpler sugars first: glucose, fructose, sucrose, maltose and then maltotriose."

In the natural environment, yeast forms buds and reproduces. But in the absence of oxygen, yeast cells can't reproduce, which is a positive in controlling the taste of beer. "Brewers of the past worked diligently, selecting and reusing yeast to the point that brewer's yeast eventually lost the ability to form spores and lost the ability to mate,"

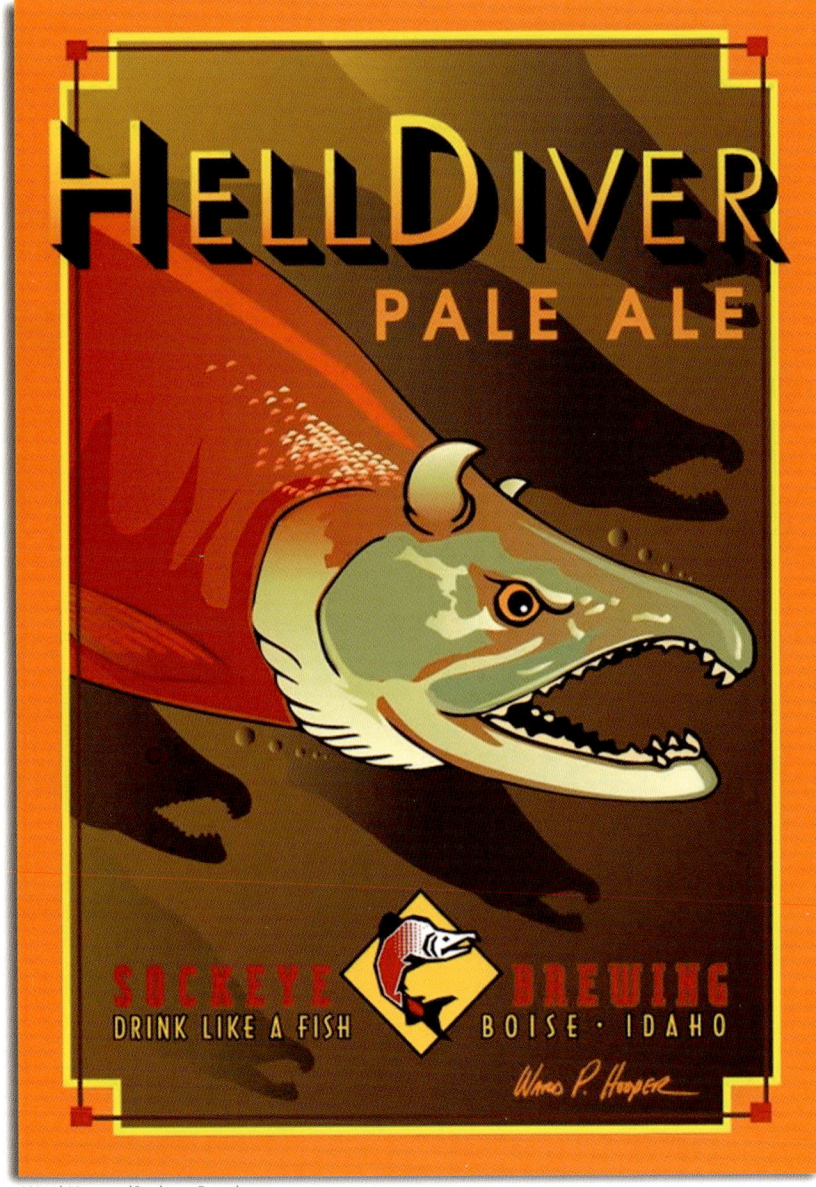

Ward Hooper/Sockeye Brewing

White and Zainasheff said. That way, "brewers can count on yeast to be more consistent from batch to batch."

The fermentation process takes several weeks, until the yeast consumes all of the sugars in the wort and drops to the bottom of the beer batch. Brewmasters check fluid density and pH levels to know when the fermentation is complete. They capture the yeast from the bottom of the tank and recycle it for the next batch.

Yeast needs 100 percent of its essential vitamins and minerals to make it through fermentation "properly nourished" and to be ready to work again in the next batch, the authors wrote in the *Yeast* book.

To complete the beer-making process, brewmasters inject just the right amount of carbon dioxide gas to the beer batch to create the fizz.

Mitch Steele, head brewer of Stone Brewing Company in California, who penned the foreword to the *Yeast* book, had a conversation with brewers at beer-making giant Anheuser-Busch about the importance of

Left:
Sockeye Brewing of Boise taps into Idaho's reverence for wild fish and powerful rivers. Brand names like Hell-Diver Pale Ale and Dagger Falls IPA honor the Salmon River of central Idaho.

Below:
Rhizopus is a genus of fungi that, like brewer's yeast, metabolizes sugar. Some species cause serious infections in humans. Bottom: Microbrewing begins with yeast packets or vials.

yeast in making beer. "Realistically, yeast can be the most active flavor ingredient in the brewing process, and it is certainly the most temperamental ingredient in beer. As any brewer knows, you must treat your yeast with the utmost care, or the beer can end up tasting horrible."

The authors encourage beer-makers to master the art of pitching and caring for yeast. "Become a yeast whisperer," they say.

Because of the influence that yeast has on the flavor of beer, it's easy to imagine how important it would be to select the right strain for making a particular type of beer. Overall, there are more than 500 species of yeast, and within each species, there are thousands of different yeast strains. Over time, beer brewers have refined their search to two main types of yeasts for making either lagers or ales—*Saccharomyces cerevisiae* (ale yeast) or *Saccharomyces pastorianus* (lager yeast). *Saccharomyces* means "sugar fungus" in Latinized Greek.

Lager yeasts are known as *bottom-fermenting* yeasts and ales are known as *top-fermenting* yeasts. During the fermentation process, ale yeasts bunch together and form a continuous, thick layer on top of the wort. This dynamic is known as *flocculation*.

In enclosed fermenters, you can't see flocculation happen. When making a batch of Udaho Gold, Matt Ganz of Salmon River Brewery showed me how flocculation looks in some open fermenters in the back room. He lifts the lid, and I see a light brown, super-thick layer of yeast smothering the top of the batch. It almost looks like that rich, brown sugar frosting you see on the top of German chocolate cake, magnified by 100 times.

magic of yeast 105

To get started in selecting a yeast strain, most brewmasters buy specific strains of yeast from either Wyeast Laboratories in Odell, Oregon, or White Labs in San Diego and Davis, California, and Boulder, Colorado. In the yeast catalogs provided by both, yeast strains are broken down into categories for ales, lagers and Belgian lagers. Brewmasters select a yeast, make a batch of beer, check on the taste, and determine if they like the result.

Francis of Payette Brewing Company selected a London ale yeast for

Left:
Idaho boom towns had more breweries than churches in the era before pressurized bottles and pasteurization. Pictured: hoisting whiskey and beers at Christie's Saloon in Troy, Idaho, about 1900; saloon token, about 1905.

Below:
Craft beermaking has exploded with the movement toward local foods. In 2014, the Brewers Association ranked Idaho ninth in the nation in breweries per capita. More than a dozen sell beer in the Boise Valley.

the pale ales he wanted to make. "I decided what kind of beer I wanted to make and went from there," he says. "I wanted to try something different, but I wasn't sure exactly what I needed. It's not a 'one and done' type of thing."

To buy enough liquid yeast for one batch of beer costs $400, Francis says. The yeast is shipped to the brewery in a plastic container. Brewers can recycle the yeast for subsequent batches over and over again, often storing the liquid yeast in small kegs.

At Salmon River Brewery, Ganz says they like to keep their selection of yeasts a secret. But he confides that he purchased an English strain of yeast for the Udaho Gold ale, and when they made the first batch, "everyone loved it."

A year after the Salmon River Brewery had been in business, a man named Adolphus Busch IV (related to THE Busch family) stopped by the brewery for lunch. He ordered a glass of Udaho Gold to go with it. He had tried Udaho Gold at Toll Station Pizza in McCall, and really liked it. "It caught my attention," Busch told the *Idaho Statesman*. "I had it a few more times and pretty much thought it was the best beer I'd had in a long, long time."

It turns out that river guide David Bingaman had taken Busch rafting on the Salmon River and talked about the popular Salmon River Brewery. (Busch has a second home in Grangeville, Idaho.) Bingaman is a big fan of Udaho Gold and the brewery. "You should go try that beer," he told Busch.

Busch returned to the brewery and talked to Ganz and his partner Matt Hurlbutt about their future plans. They ended up working on a business plan together for two years before all the kinks were worked out, Ganz says. Ultimately, Busch bought 49 percent of the business to allow Salmon River Brewery to expand into the retail market

Fermentation

Fermentation spoils and elsewhere preserves. It sours and sweetens, infects and kills infections, grows hormones and antibiotics. For cells it is a way of making energy without oxygen. For the ancients it was the magic of preserving food for storage. For sauerkraut aficionado Sandor Katz, in his popular book about fermentation, it was the flavorful space between fresh and rotten where food acquired its zest.

Bacteria and fungi of many kinds become agents of fermentation when they break down sugars and starches. *Lactobacillus* bacteria curdle milk into cheese. *Oenococcus oeni,* a bacterium, softens the texture and adds complexity to wines such as Chardonnay. *Penicillium notatum*, a mold fungus, mushrooms the Earth's population by combating infectious disease.

Pictured: Cork-popping champagne is twice fermented—once in the vat and again in the bottle. Insets: Hieroglyphs suggest that fermented beverages were common in ancient Egypt by the third millennium BC; pharmaceutical fermentation; making penicillin in the 1950s; budding yeast cells.

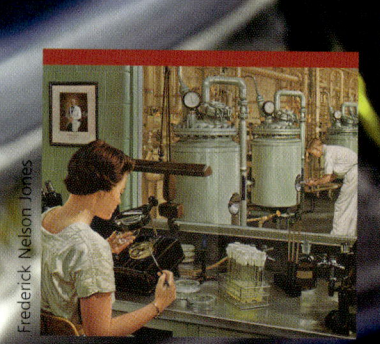

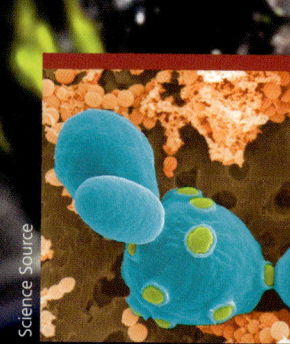

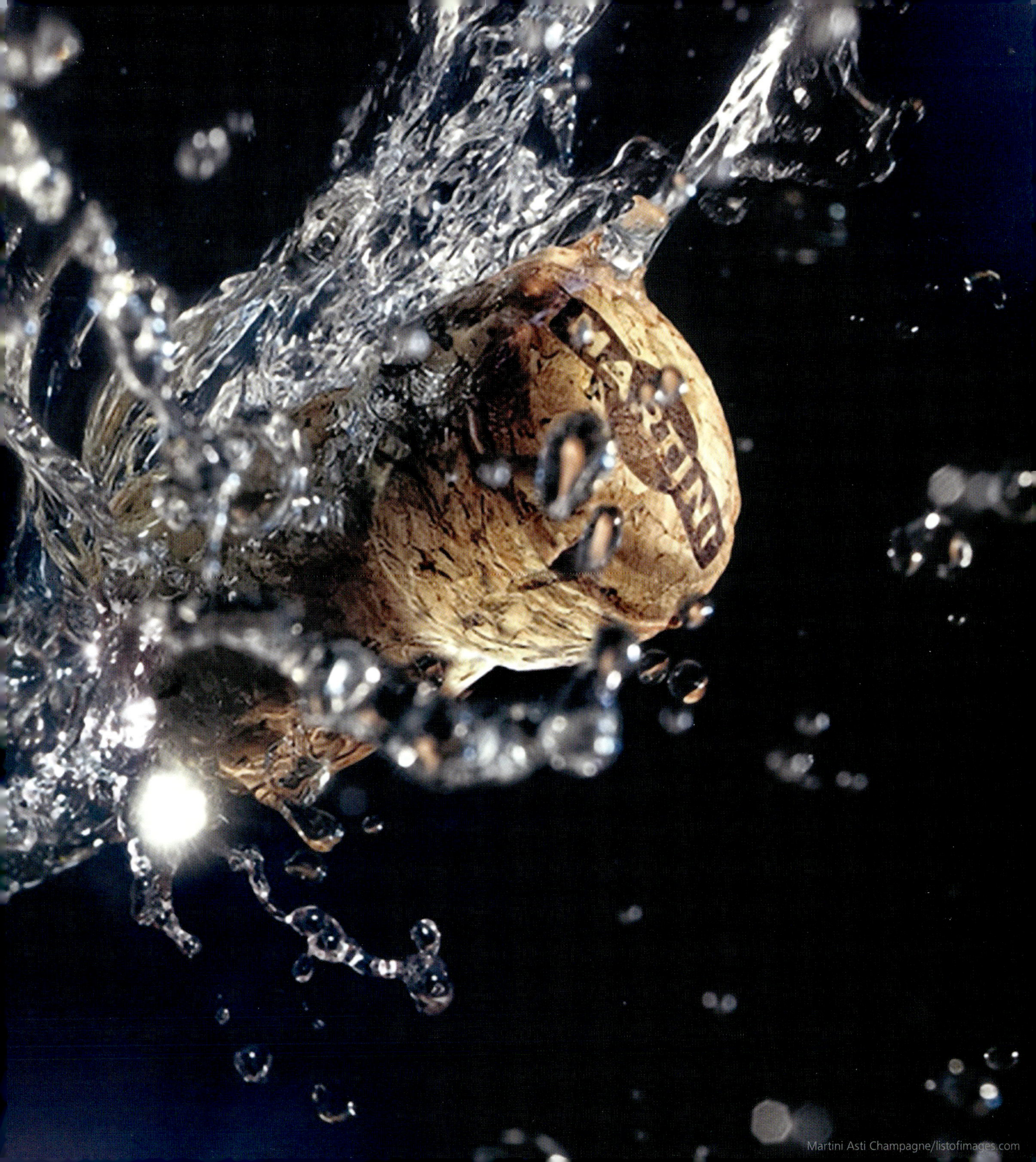
Martini Asti Champagne/listofimages.com

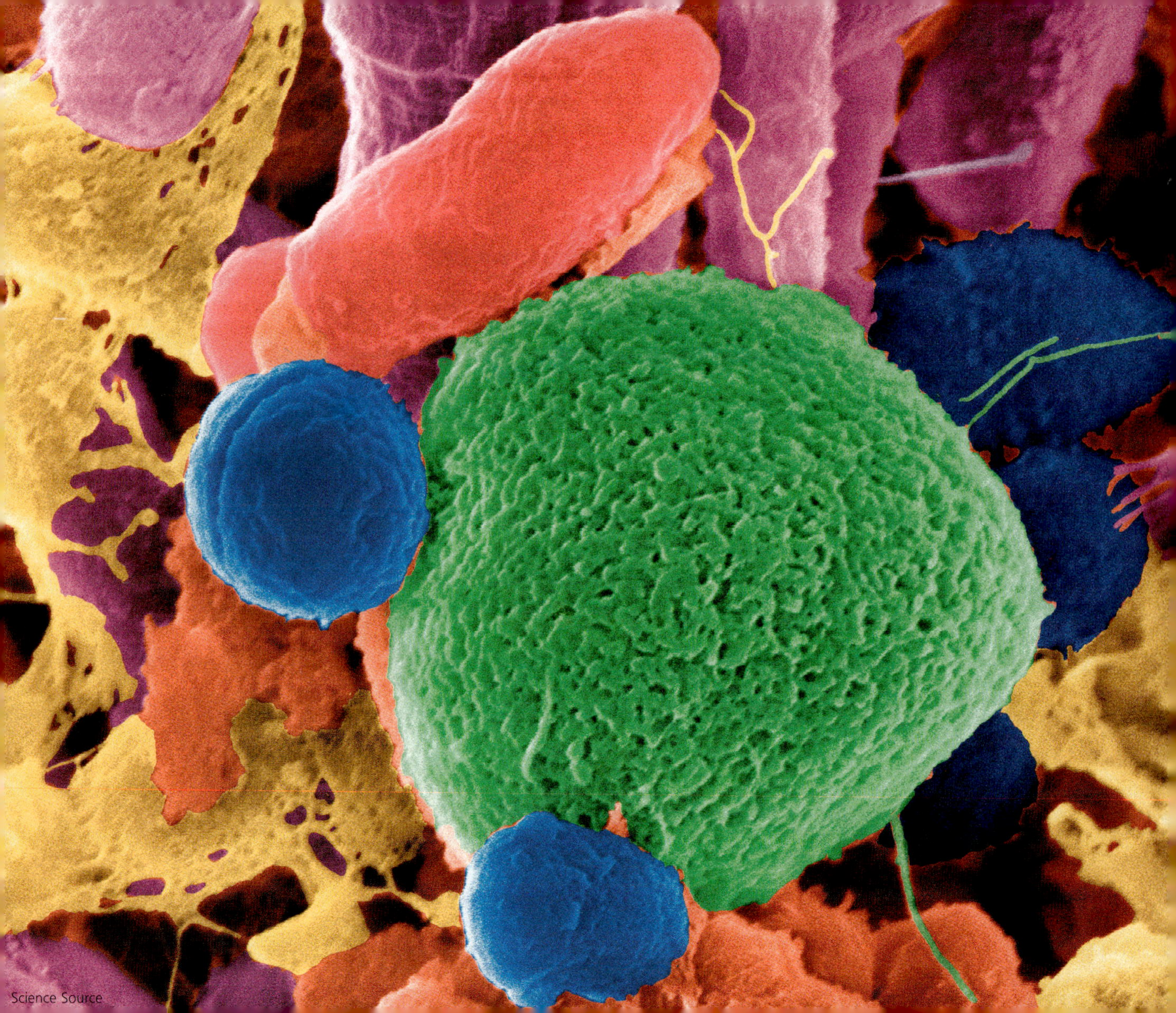
Science Source

don't be fooled!

giardia lurks in

pristine mountain streams

5

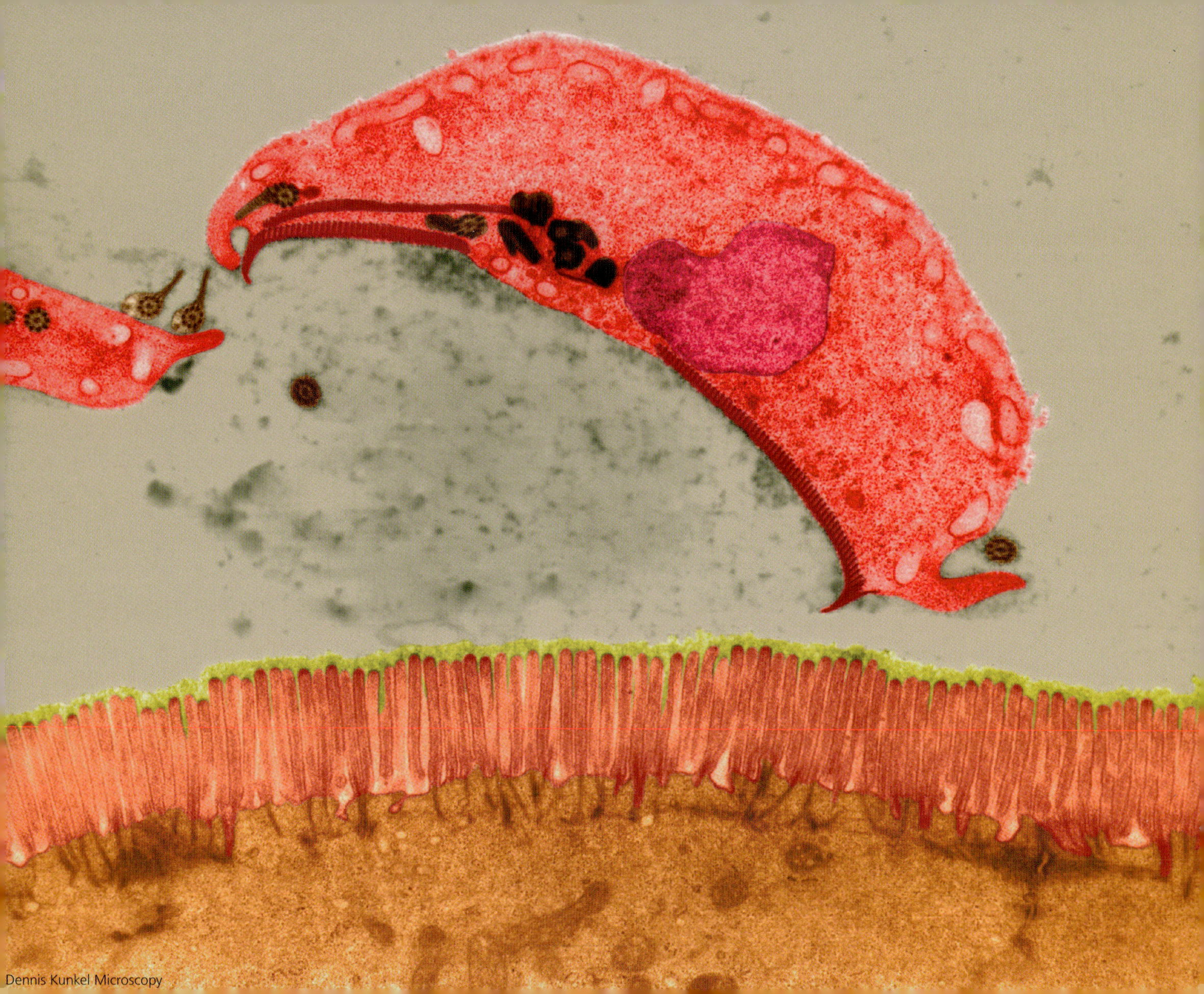
Dennis Kunkel Microscopy

Left:
Giardia intestinalis, also called *Giardia lambia*, colonizes and reproduces in the upper part of the small intestine. Globally the parasite afflicts some 2.8 million people each year.

Previous:
Giardia enters the human body in its dormant form as a cyst. The cysts can be hardy enough to survive months of sub-zero cold and even swimming pool chlorination.

On the morning of Sunday, May 18, 1980, Mount St. Helens erupted in central Washington, sending a huge mushroom-shaped plume of ash into the sky. Little did I know that the volcanic eruption would inexplicably introduce me to a little-known microbe that forever changed the way that I—and countless other backpackers—would experience the wilderness in the Pacific Northwest.

On that Sunday, I was riding the final day of a two-day, 230-mile bicycle tour called the Tour of the Swan River Valley in western Montana. Fortunately, we were more than 700 miles away from the volcanic eruption. I was a junior at the University of Montana at the time. Picture a lanky young man with bushy brown hair and a beard, wearing shorts and a T-shirt, with Converse high tops strapped into the toe clips.

By about 2:30 p.m. that afternoon, gray clouds turned jet black on the western horizon as the ash plume neared Missoula. Pretty soon, the ash started falling like powder snow. It was surreal. We were poor college students—we didn't even have a TV—so we had no idea what was going on. But some neighbors told us what had happened. The volcanic eruption was the most deadly in American history—more than 50 people had been killed, and 200 homes, 185 miles of highway, 27 bridges and 15 miles of railroad had been destroyed.

By 8 p.m. that night, Missoula was covered with several inches of volcanic ash. Health authorities recommended that people should stay inside. University of Montana officials closed the campus for a week. Hooray! That meant we had an extra week to get ready for finals. But instead of cocooning indoors, several friends and I decided we should get the heck out of town and escape the ashen zone. We planned a backpacking trip to the Absaroka Wilderness northeast of Yellowstone National Park. Our game plan was to

Below:
Giardiasis by state per 100,000 population, 2010. New England suffers the most. Idaho and Colorado lead the American West.

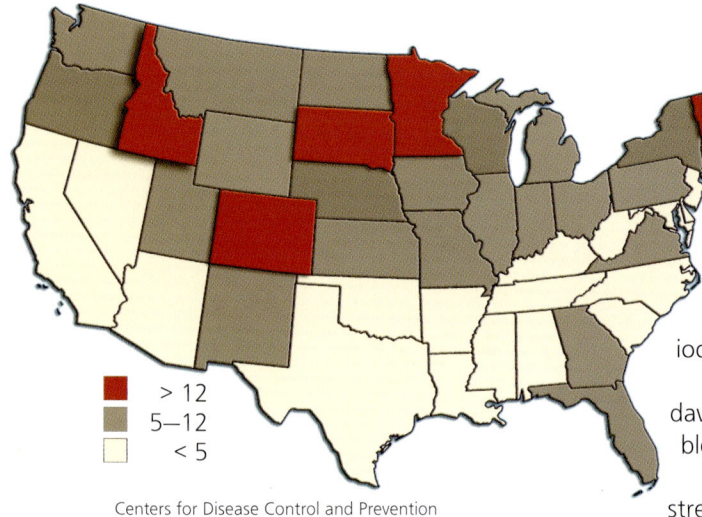

> 12
5–12
< 5

Centers for Disease Control and Prevention

hike into the woods and see how far we could get before we ran into snow.

Back in those days, when it came time to fill the plastic water bottle with drinking water, we'd dip it into the creek and fill it up with the seemingly pure mountain water. When it came time to cook, we filled the cook pot directly from the creek. Never thought twice about it. No one had a water filter or iodine tablets.

The morning of Day 3 in the Absaroka Wilderness dawned with gray, overcast skies. I woke up feeling horrible. I had diarrhea, and I couldn't eat breakfast.

Somehow, I had to rally. I barely had enough strength to wobble as if I were a zombie. Then I'd feel sick, and I'd quickly step off the trail to vomit. That happened multiple times. Somehow, I got out of there and collapsed in the back of the truck.

Ten days later my stomach began to make loud noises and I got a bad case of stomach cramps. The doctor informed me that I probably had giardiasis (commonly referred to as giardia). I had never heard of it. He asked me for a stool sample and explained that *Giardia* was a single-celled parasite that could be ingested by drinking water directly from mountain streams, that is, without treating the water first. Because of the uncontrollable nature of my bowels, he took pity on me and prescribed Flagyl, an antibiotic. Within 36 hours, I was feeling much better. Thank goodness! The full antibiotic treatment took a little over a week.

I took a keen interest in *Giardia intestinalis* following that experience 30 years ago. A single-celled protozoan parasite, *Giardia intestinalis* is a troublemaker on a global scale—about 280 million cases of giardiasis are reported

Left:
W.T. Rawleigh Company, founded in 1889, marketed a snake-oil medicinal cure-all for stomach ailments such as cholera and giardiasis. The mixture was 48 percent alcohol.

Below:
Atabrine, a trade name for quinacrine hydrochloride, became the WWII drug of choice for malaria and giardiasis. Pictured: gruesome warning at an army field hospital in Papua, New Guinea, 1943.

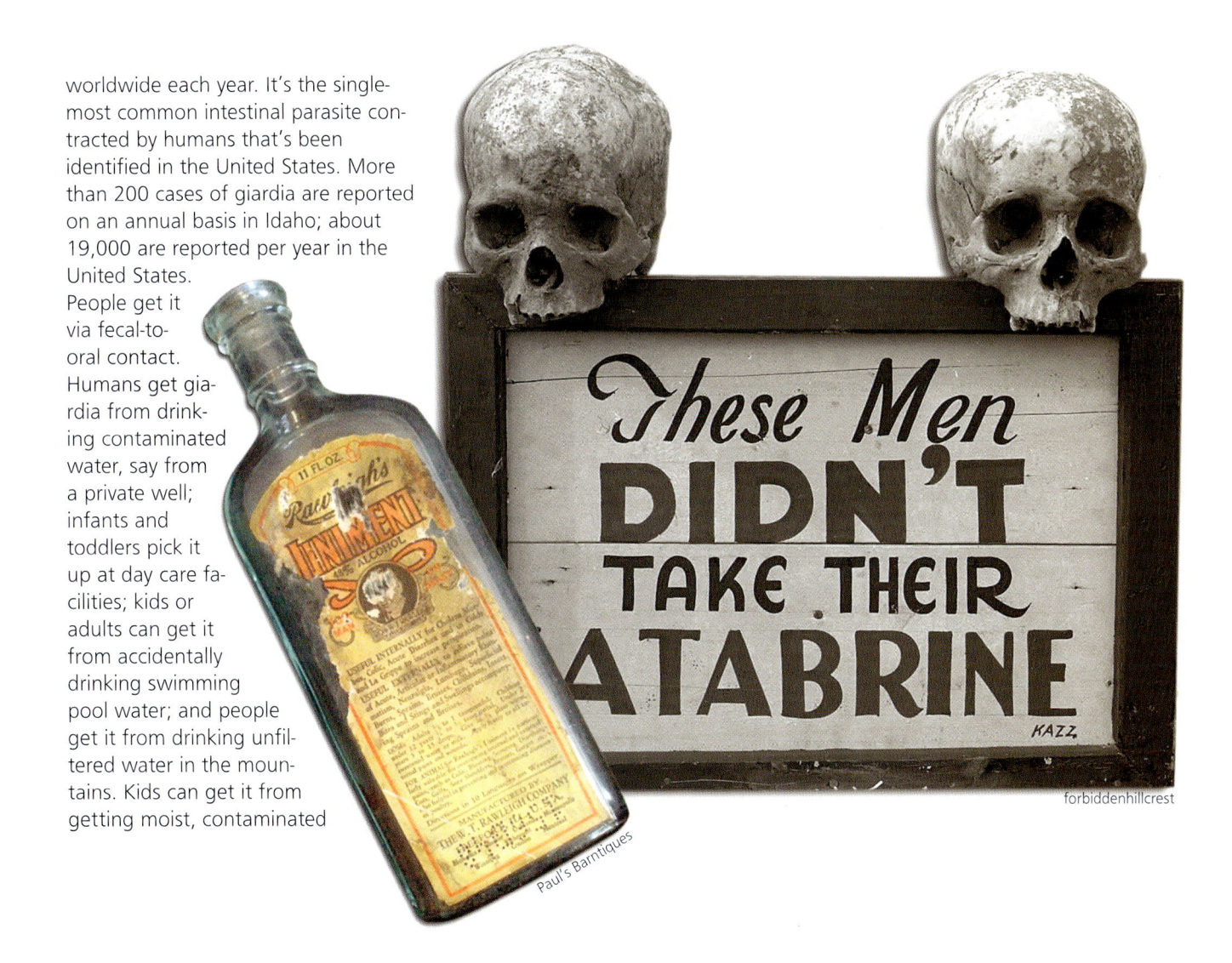

worldwide each year. It's the single-most common intestinal parasite contracted by humans that's been identified in the United States. More than 200 cases of giardia are reported on an annual basis in Idaho; about 19,000 are reported per year in the United States. People get it via fecal-to-oral contact. Humans get giardia from drinking contaminated water, say from a private well; infants and toddlers pick it up at day care facilities; kids or adults can get it from accidentally drinking swimming pool water; and people get it from drinking unfiltered water in the mountains. Kids can get it from getting moist, contaminated

giardia lurks 117

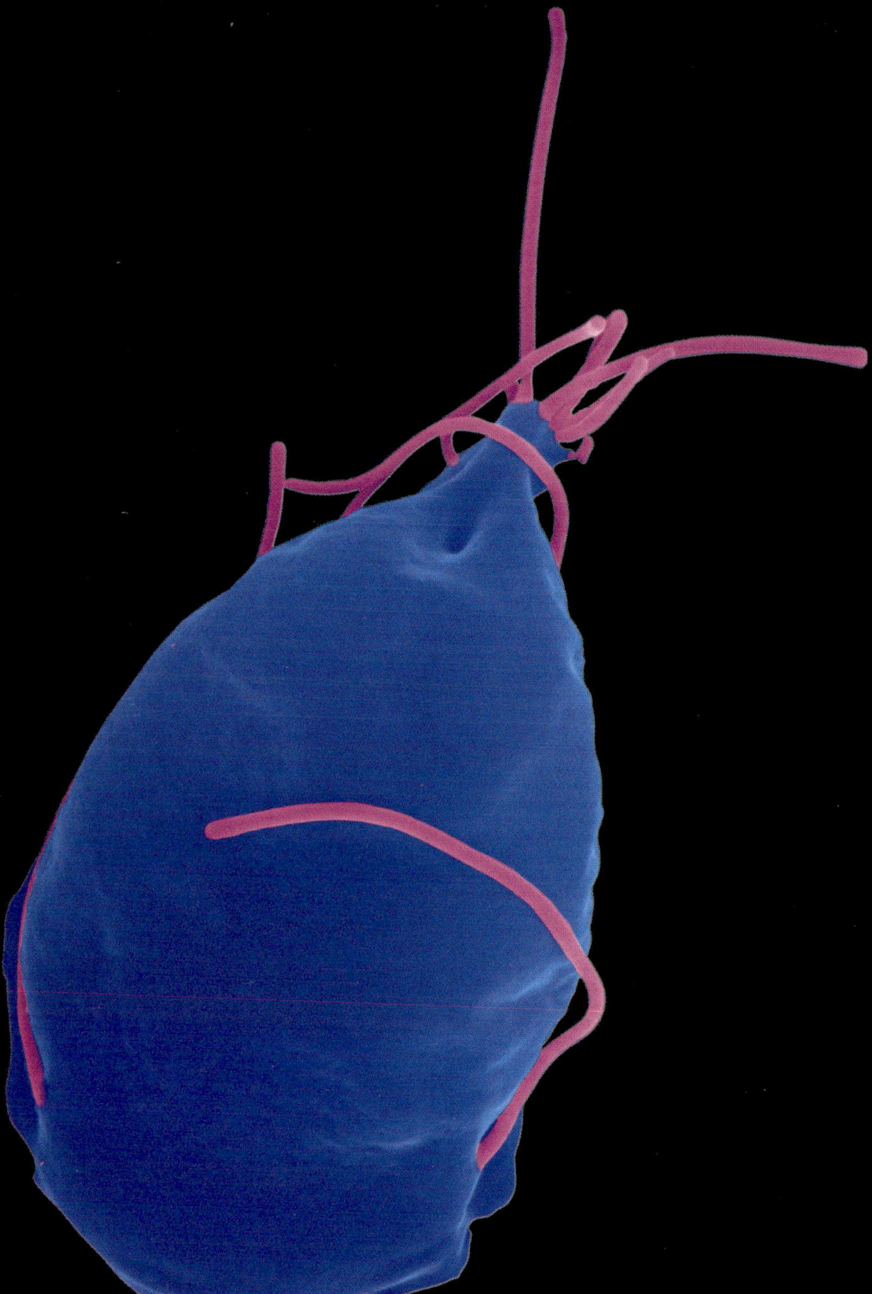

Dennis Kunkel Microscopy

Left:
Eight whiplike flagella propel the adult parasite through the host. Zoologists have identified at least 30 types of *Giardia* and six morphologically distinct species.

Below:
Giardia has two morphological stages: cyst and trophozoite. Cysts, when ingested by birds and mammals, morph into free-floating trophozoites.

dirt in their mouth. In more obscure examples, couples can ingest *Giardia* from oral-anal contact, mountain climbers have ingested *Giardia* from glacial runoff, and unsuspecting whitewater rafters have gotten it by unwittingly swallowing river water after a raft flips in a big rapid.

Giardia cysts are so tiny, they're not visible to the naked eye. They measure 5 to 15 micrograms in width and 8 to 18 micrograms in length. "There's a response curve: the more cysts you ingest, the higher the likelihood you'll get infected," says Jonathan Yoder, an expert in foodborne, waterborne and environmental diseases for the Centers for Disease Control and Prevention in Atlanta, Georgia. "The highest level of contamination would be the highest risk. People do become ill after drinking from streams that look really pure and clean. That's your unlucky day."

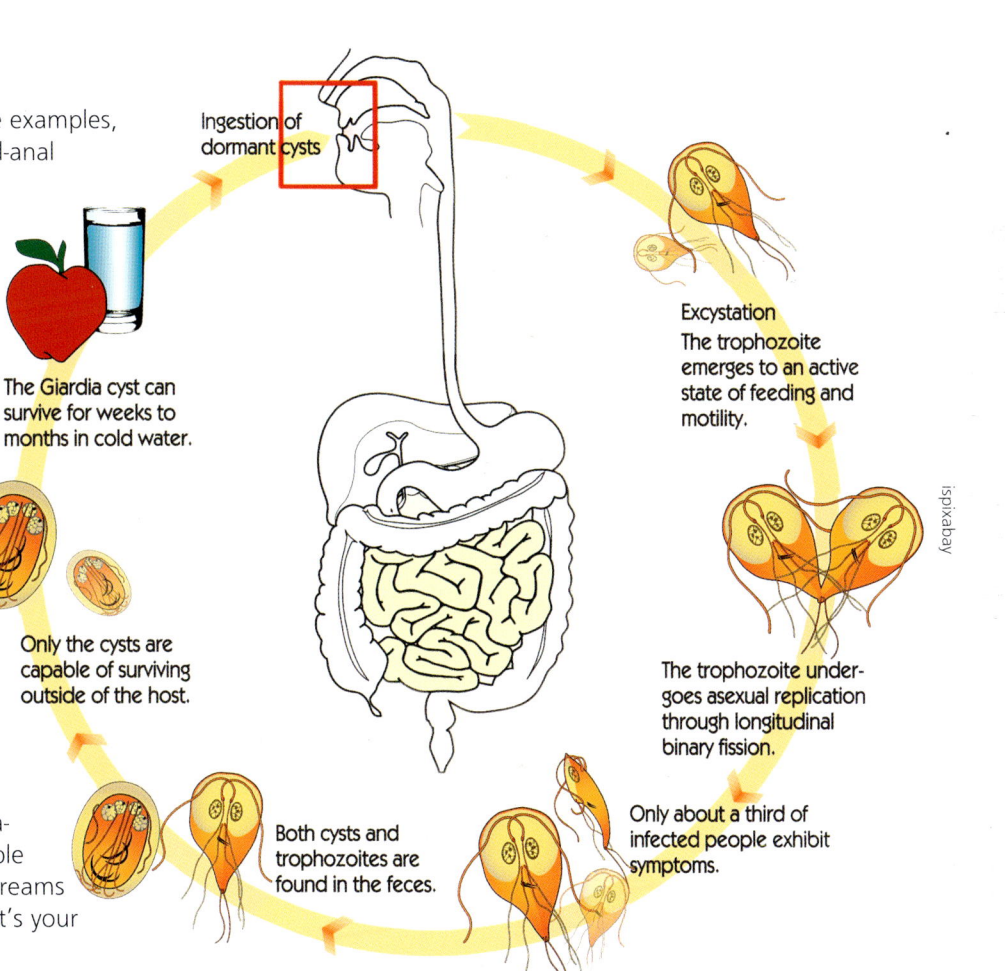

Ingestion of dormant cysts

The *Giardia* cyst can survive for weeks to months in cold water.

Only the cysts are capable of surviving outside of the host.

Both cysts and trophozoites are found in the feces.

Only about a third of infected people exhibit symptoms.

The trophozoite undergoes asexual replication through longitudinal binary fission.

Excystation
The trophozoite emerges to an active state of feeding and motility.

ispixabay

giardia lurks 119

Below:
King Louis IX (1214—1270) died during an outbreak of dysentery after landing with his army in Carthage during the Eighth Crusade. Historians suspect giardiasis.

When I started digging into the topic for this book, I wondered how long *Giardia* has been present on planet Earth. In 1681, Antonie van Leeuwenhoek was the first to document *Giardia*, which he saw when examining his stools under his pioneering microscope. King Louis IX, the leader of the Seventh Crusade, may have suffered from giardiasis in the mid-1200s. The king had "such serious diarrhea that part of the

Top:
Alfred Giard (1846–1908) gave his name to the *Giardia* genus. A Darwinian, the French zoologist also endorsed Jean-Baptiste Lamarck's rival theory of inherited characteristics.

Bottom:
Giardiasis was commonly known as beaver fever during the heyday of the Idaho fur trade. Sacajawea, trekking west with Lewis and Clark, suffered the gastrointestinal "fever" in 1805.

monarch's breeches were cut away to ease his personal hygiene," according to a science article in *The New York Times*. An archaeological dig discovered the presence of *Giardia* in a medieval latrine used by the king, according to the *Journal of Archaeological Science*.

When did it come to the United States, or was it here all along as part of the natural order of things? Experts say it's impossible to know for sure, but most likely, giardia would have caused grief for the Lewis and Clark Corps of Discovery in the early 1800s. "It's just speculation on my part, but my guess is that it was present in different continents at that time," Yoder says.

So how did it get there? *Giardia intestinalis* can be carried in cyst form by humans, mammals and birds, and it's spread into the environment via their feces. Giardia has been referred to in slang as "beaver fever." Mountain men and trappers may have come up with that moniker. Because beavers can carry *Giardia* cysts, they could have infected many of the major river systems in the United States. Beavers were numerous in the Rocky Mountain West in the early 1800s, when mountain men followed Lewis and Clark into the region. During the fur trapping era (1812–1830), "beavers were as thick as fleas on a dog," wrote Carl Burger in the book *Beaver Skins and Mountain Men*.

Once a stream becomes contaminated with *Giardia*, the hardened *Giardia* cysts can persist in cold water for weeks, if not months at a time, according to health experts. For an area to become contaminated on a consistent basis, there would need to be multiple hosts living in the area, such as bears, rodents, elk, deer, beaver, and even songbirds, all of which could repeatedly contaminate water bodies by shedding the *Giardia* parasite through their feces. Add human carriers and dogs, and one can see

Dennis Kunkel Microscopy

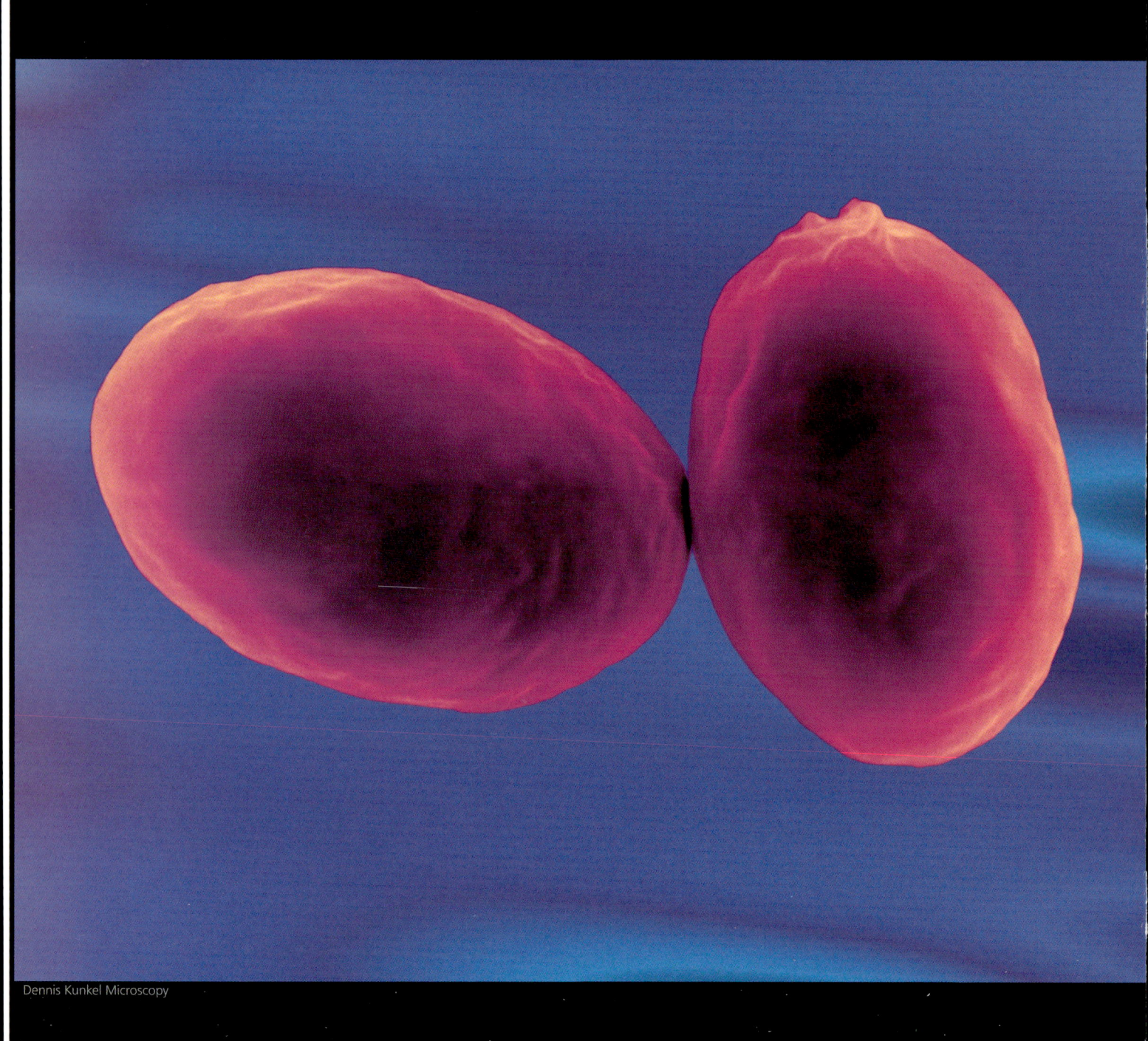

Left:
Cysts can be large enough to be trapped by water filters with a pore size of one micrometer or less. Cysts also can be killed by iodine, halazone and other disinfectants.

Below:
Hookworms chew into the wall of the intestine, drinking the blood of their hosts. Other parasitic worms common in untreated water are tapeworm, roundworm and pinworm.

trophozoites reproduce asexually through a process of binary fission, creating havoc in the small intestines. This is what causes discomfort in humans, when cramps and loose bowels begin.

Elizabeth "Lee" Hannah, an associate professor of epidemiology at Boise State University, says stomach acid normally can neutralize various forms of bacteria and parasites, but the hardened *Giardia* cysts pass through that stage without being affected. Once they enter the small intestine, the *Giardia* cysts somehow know they have "entered this warm moist place" where the environment is perfect for the trophozoites to multiply, Hannah says.

Who knows how many trophozoites are produced during the colonization of the small intestine, she says, but "I've looked at *Giardia* stool samples from dogs, and I've seen over a hundred on one specimen slide," she says.

The human body senses that something is going wrong in the small intestine, Hannah says, and it produces more body water into that area to dilute and wash out the trophozoites. When *Giardia* trophozoites move toward the colon, they retreat into cyst form, and they're discharged from the body in feces.

For many people, *Giardia intestinalis* can be warded off with time without medication, according to the Centers for Disease Control and Prevention. But for more severe cases, such as the one I had, medication is needed. In some cases, people can end up with irritable bowel syndrome, pronounced arthritis and other long-term consequences.

On a national level, Yoder says the most popular way that people get giardia are via:

1. Person-to-person contact. Kids in day care centers, who may have poor hygiene, spread it from touching feces.
2. Drinking contaminated water from sources that are supposed to be clean. Private groundwater wells often are the culprit. Some municipalities treat

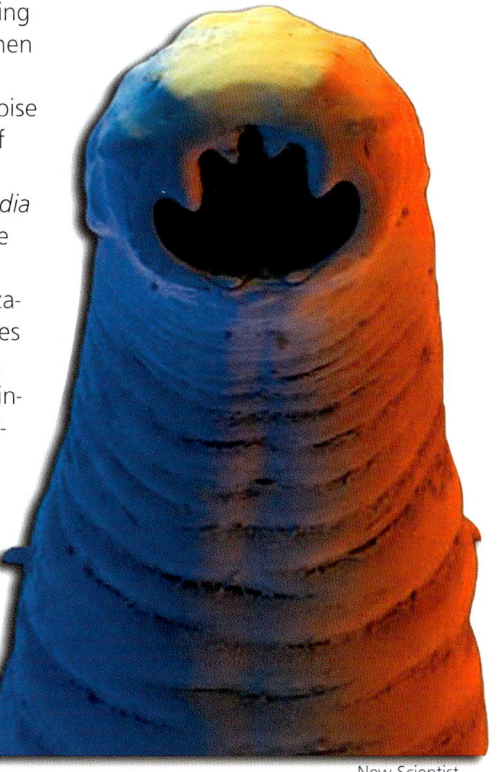

New Scientist

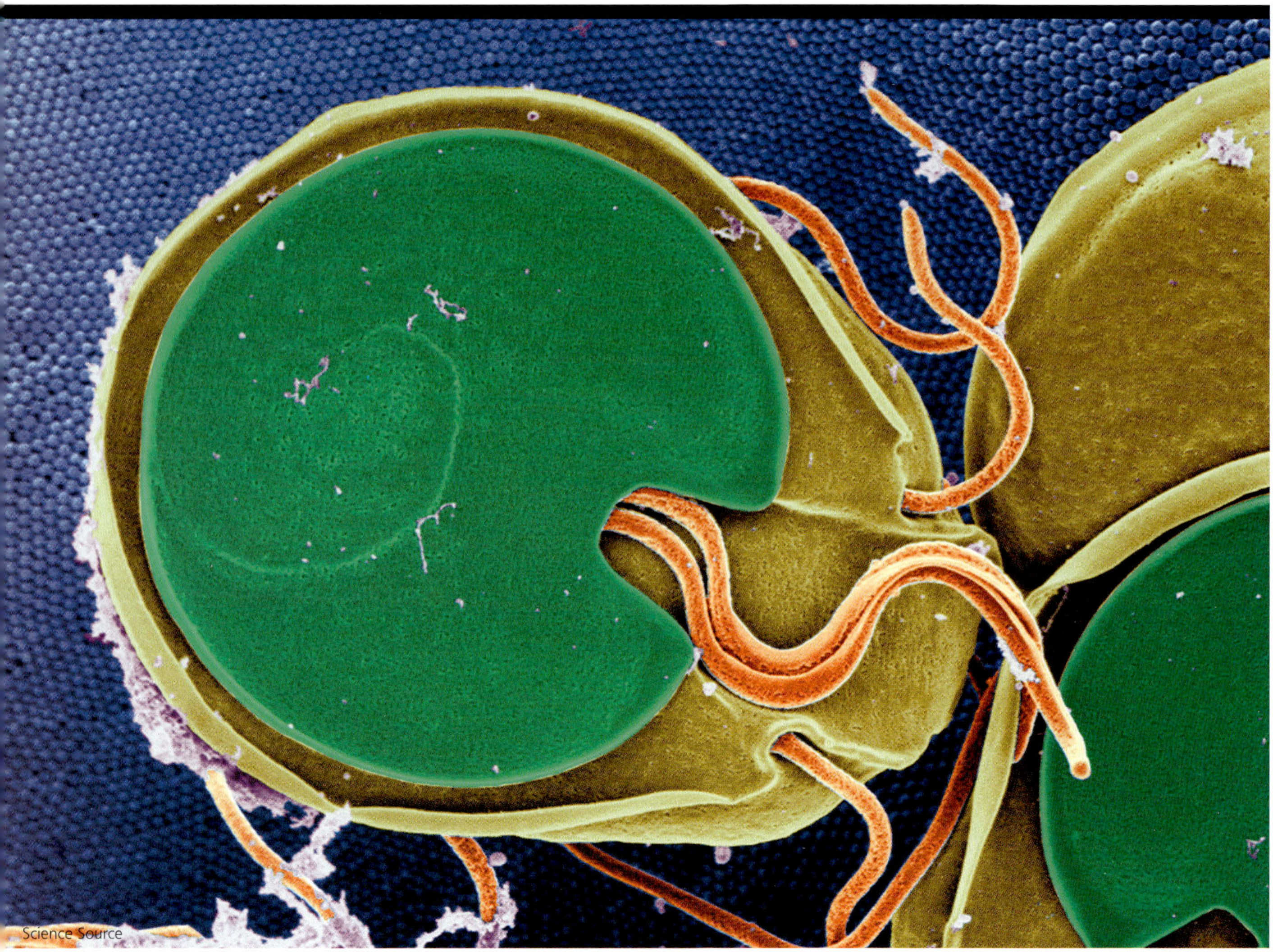

Science Source

Left:
Each egg-shaped *Giardia* cyst can reproduce and divide into eight hungry trophozoites. Ten cysts can be infectious enough to painfully cripple the host.

Below:
Vomiting and watery-green diarrhea are common giardiasis symptoms in dogs. Metronidazole tablets are an effective treatment.

water with chlorine and sophisticated filtration systems that not only catch giardia but also *Cryptosporidium*, a smaller microorganism that's more commonly found in water bodies near civilization. It's usually carried by cattle, sheep and goats, particularly young, pre-weaned animals. The city of Milwaukee, for example, experienced an outbreak of *Cryptosporidium* that affected 400,000 people in 1993. The city's drinking water came from Lake Michigan, but the filtration systems were not able to sufficiently catch the tiny microscopic creature at the time.

3. Any type of direct fecal-oral transmission, such as during sex.
4. People accidentally ingesting *Giardia* from recreational water sources such as lakes, streams and swimming pools that haven't been adequately maintained with chlorine and filtration.

In Idaho, the same primary sources apply, but because it is a mountain state with abundant public forests and desert lands, Hannah postulates that we have a higher number of people who are active in outdoor recreation, and thus, a comparatively larger number of giardia cases from people who are involved in backcountry recreation.

"A lot of people who move to Idaho tend to be outdoor enthusiasts," she says. "We spend more time fishing, boating, hiking, camping and backpacking. The Pacific Northwest in general has a higher prevalence of these kinds of activities, so it makes sense that we are going to have more people who get giardia while engaged in mountain recreation."

perroswiki.blogspot.com

askbabycotswold.co.uk

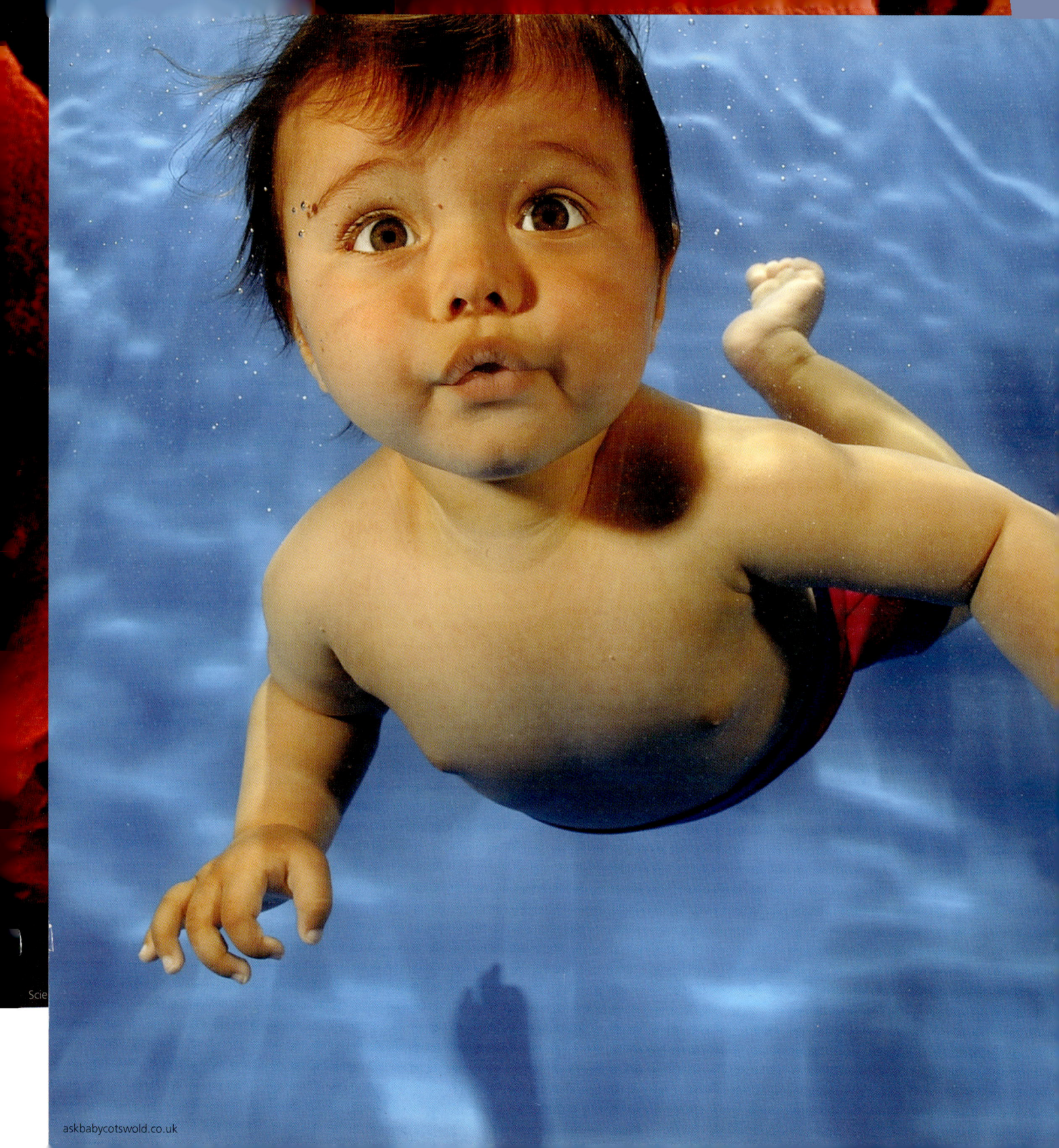

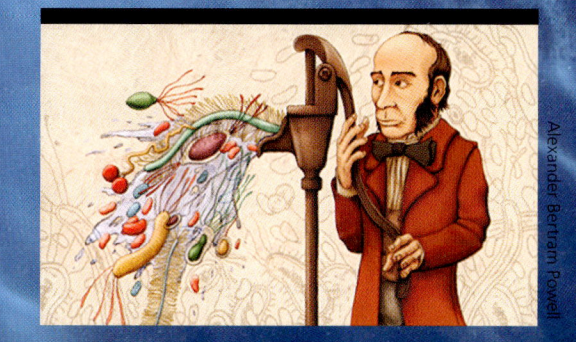

Thirsting for Drinkable Water

The United States is fortunate to have engineered some of the world's safest and most complex water delivery systems. American cities mostly use chlorine disinfectants to kill pathogenic microbes. Eight of ten Americans drink chlorinated water from municipal treatment plants.

The thirsting for community water systems dates back at least 6,000 years. Greeks of ancient times kept records on boiling, straining, and charcoal filtration. Roman cities used complex drains and aqueducts to separate water from waste.

British physician John Snow was the first to convincingly show that bad water spreads pathogens. In 1854, while cholera killed tens of thousands, Snow disabled a pump at an infected public well on Broad Street in London's Soho. Quickly the epidemic subsided. It remained for biologists Robert Koch and Louis Pasteur to explain how water transmitted cholera germs.

Municipalities have chemically treated for dangerous bacteria and protozoa since Chicago and Jersey City, in 1908, pioneered chlorination. Today the U.S. Environmental Protection Agency sets treatment standards that guard against pathogens. Pictured: John Snow pumps London's microbes, 1854. Left: *Vibrio cholerae*, still the scourge of the impoverished. Below: Country-by-country share of the world's population with access to treated drinking water, 2006.

Share of population with treated water
- 85% or more
- 70–85
- 55–70
- 40–55
- less than 40%
- No data

World Health Organization/UNICEF

Government of Alberta

forest epidemic

blister rust fungus invades the Pacific Northwest

6

Chi-E-Shenam Westin

Left:
Blister rust (*Cronartium ribicola*) infects tree tissue in the purplish stain of a microscope slide. The fungus, native to Asia, reached the Rocky Mountains via seedlings imported from France.

Previous:
A Canadian fire crew burns rust-prone shrubs in a forested understory of parasites and pathogens. Bottom: Rust fungi release their windblown spores through hornlike protrusions called *telia*.

A brisk wind rages across the top of Brundage Mountain near McCall on a clear, mid-October day. It whips through dozens of gnarled whitebark pine trees that punctuate the upper story of the 7,640-foot mountain, causing a chorus of eerie, high-pitched sounds suitable as a sound track for a horror movie. Looking around, it's hard not to notice how many of the whitebark pines are dead. Especially on this windy day, we might call them "gray ghosts," considering the specter of what may happen to these ancient trees that lord over the tops of mountains and ski areas in central Idaho and the Northern Rocky Mountains. Some of the whitebark pines are more than 800 years old—far older than any other trees in the forest.

Healthy whitebark pine trees have been under siege in the past decade by a major mountain pine beetle outbreak. A native insect, mountain pine beetles normally prey on lodgepole pine trees and decimate large stands of them, which they have done in the past decade in the Stanley area and large portions of the Salmon-Challis National Forest in east-central Idaho. But hotter temperatures brought on by global warming have allowed mountain pine beetles to reach into higher elevations than the winged bugs normally go and prey on whitebark pine trees.

The result is that whitebark pine trees have been killed by mountain pine beetles at alarmingly high rates. Mortality has ranged from 30 percent to 97 percent in 42 whitebark pine stands in Idaho, Montana and Wyoming. The density of live whitebark pine trees dropped by more than 80 percent on more than half the areas surveyed. The good news is that the mountain pine beetle epidemic is over for now—bark-beetle outbreaks usually last about a decade.

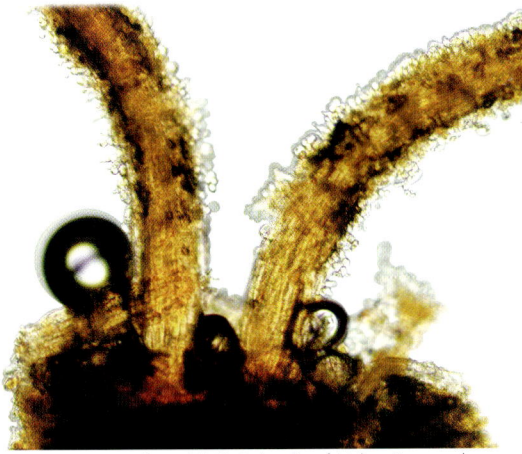

Bruce Watt, University of Maine, Bugwood.org

Mendocino Historical Society

Left:

Lumber baron Frederick Weyerhaeuser led the turn-of-the-century rush to the evergreen pines of Idaho's Clearwater basin. From 1900 to 1927, Weyerhaeuser acquired 220,000 acres with 2 billion board feet of western white pines. Pictured: logging Idaho, about 1900.

Below:

Flowering species of the *Ribes* genus offer a leafy platform for the launching of blister rust spores. Since the fungus cannot pass directly from pine to pine, ribes shrubs provide an intermediate host. Pictured: vintage poster of red currants (*Ribes rubrum*), 1894.

"There was definitely a lot of whitebark pine killed in this last outbreak of mountain pine beetles," says Laura Lazarus, an entomologist who works on insect and disease issues in Idaho's national forests south of the Salmon River.

But there is another, equally large threat that has been killing whitebark pine trees for decades. White pine blister rust, *Cronartium ribicola*, an exotic fungus native to Asia, was unwittingly imported into the Pacific Northwest from white pine from seedlings grown in France in 1910. It has been preying on whitebark pine trees—as well as white pine, Idaho's state tree—since that time. Mountain pine beetle epidemics in the 1930s, in the 1970s and from 2002 to 2012 created a one-two punch that has killed an unprecedented number of whitebark pine trees and raised concern about whether the species can survive.

Beginning in the early 2000s, as the latest mountain pine beetle outbreak started its rapid advance, forest pest experts sounded the alarm about the mortality levels being observed in whitebark pine. The trees, known as a keystone species of high-elevation ecosystems, are important from a watershed perspective because they hold the soil together in areas with deep snow and heavy runoff. Live trees help retain snowpack by providing shade. The large, highly nutritious seeds produced by whitebark pines are key to the diets of Clark's nutcrackers, black bears, grizzly bears, red squirrels and more. By the mid-2000s, U.S. Forest Service officials crafted forest management strategies and recommendations to stem the tide. In the meantime, attorneys with the Natural Resources Defense Council (NRDC) filed a petition with the U.S. Fish and Wildlife Service to designate whitebark pine as an endangered species. In 2011, the Fish and Wildlife Service ruled that the species is "warranted" for listing but "precluded" because of other priorities. However, it is now considered a candidate species of concern.

Bibliographisches Institute Leipzig

country skiers on the lee side. Almost every ski area in Idaho has whitebark pine trees growing near the summit. The trees tend to grow above 7,000 feet in elevation to tree line at 10,000 feet. They are not particularly tall trees, often barely 100 feet high, but they are stocky and sturdy with large-diameter, twisted trunks to survive in high-elevation mountain areas. Even dead whitebark pines remain standing for decades, holding the soil together with their extensive root systems.

The Clark's nutcracker, a gray songbird with black wings and white tail feathers, plays a crucial role in the natural reproduction of whitebark pine trees. The nutcracker and the whitebark pine must have evolved together because the trees are dependent on the nutcrackers to sow their seeds. Every year, the birds dutifully collect seeds from whitebark cones and cache them in the ground. Other pine species have wings on their seeds for natural dispersal by the wind.

The nutcrackers, named by explorer William Clark, have large, stout bills that allow them to extract seeds from the cones. The birds are equipped with a special pouch in their throat where they store seeds as they fly to a new cache. The special pouches can hold up to 100

NW bird blog

Left:
Whitebark pines interdepend on seed-hording birds and rodents. Clark's nutcrackers use dagger-like beaks to puncture the seedy pine cones. Red squirrels also hoard and disperse the seeds.

Below:
Black bears and grizzlies raid pine seeds buried by squirrels. The seeds, called pine nuts, are high in fat and pungent enough for a bear to sniff out through snow. Pictured: an American black bear cub.

seeds, according to experts. But get this: Individual nutcrackers may cache up to 80,000 seeds in one year. Studies show that in a year with a good cone crop, the birds will cache at least 35,000 seeds, and they will return to consume 50 percent of them. The rest are left to germinate as new trees.

"I've seen a whole bunch of nutcrackers in a meadow caching seeds at the same time," Lazarus says. "It's really cool."

Squirrels in the forest also will cache whitebark cones in debris piles called *middens*, where they can consume the seeds during the winter. In Yellowstone National Park, research shows that grizzly bears forage heavily on whitebark pine seeds, and when they have an abundant seed crop, the bears' reproductive rates will increase. Good whitebark seed supplies also will keep the bears in higher-elevation zones instead of near campgrounds, where human-bear encounters can be deadly.

Black bears also eat the seeds. Wildlife biologists in the Boise National Forest have collected bear scat that's full of whitebark pine seeds. "We see lots of black bear scat with the seeds in it," says Nadine Hergenrider, a wildlife biologist for the Lowman Ranger District. She created an educational display about whitebark pine, Clark's nutcrackers and bear scat in the entryway to the Lowman District office. "I like that the trees grow in the high country and the mutual relationship they have with Clark's nutcrackers," Hergenrider says.

Jeremy's bear blog

All told, approximately 110 species of mammals and birds depend on whitebark pine trees for survival, including the blue grouse, a large game bird that survives almost completely on pine needles in the winter.

I tag along with Forest Service entomologist Laura Lazarus and fire specialist Paul Swenson to Scott Mountain, a high peak and fire lookout in the Boise National Forest near Garden Valley, a site where the Boise National Forest has been earnestly working on restoring whitebark pine for a decade. We see quite a few whitebark pine trees with "red flags." The trees are not literally flagged with red tape. But the needles on the branches of living trees are turning a deep red rust color, showing that they are stressed and in trouble. We also see a number of new whitebark seedlings growing in the forest from restoration work.

The red flags are one telltale sign that the whitebark pines are being preyed on by white pine blister rust. Another obvious sign is that the tree may be bulging unnaturally where a tree limb joins the main stem of the tree or the bole. These are called *cankers*.

But how do the fungal pathogens reach the tree branches? How do they kill the whitebark pine?

Left:
Blister rust attacks a pine via its needles. Traveling through twigs to branches, the fungus blisters the bark with yellowish cankers. The cankers then girdle the trunk and cut the flow of fluids.

Below:
Only fragments remain of North America's ancient forest. South of Canada, at the rate of about a square mile per person since the Pilgrims landed in Plymouth, loggers have felled more than nine out of ten old-growth acres.

White pine blister rust, *Cronartium ribicola*, was imported by accident via a shipment of white pine seedlings from France to Vancouver, British Columbia. In the early 1900s, European tree nurseries often grew white pine from seed from North America, and it was shipped back with the "undetected hitchhiker," says John Schwandt. Apparently that first shipment involved 1,000 eastern white pine seedlings, many of them infected with the rust fungus.

Over a period of decades, the blister rust infected a broad range of whitebark pine in North America, and it also preyed on the stately white pine. The fungus caused huge devastation to white pine, but the impact on whitebark pine took longer to become apparent. Plus, whitebark pine has never had any commercial value, so all restoration efforts were focused on the white pine for decades.

At least mountain pine beetle outbreaks go away after 10 years—the bugs tend to eat themselves out of house and home by killing so many trees, and then the insect outbreak wanes until the next cycle. But the white pine blister rust has an amazing life cycle that perpetuates the fungus and allows it to continually attack whitebark pine trees in the high-mountain zones. Plus, as an invasive species, it can cause more harm than a native insect or disease would because the host trees don't contain natural enemies to fight off an exotic fungus that didn't evolve in the Northern Rockies.

Let's explore the life cycle of the single-celled fungus and see how it works. It all starts in the fall, when windborne basidiospores infect the needles of a whitebark pine, carried there by the wind from another host, a currant or gooseberry shrub nearby. A basidiospore is a reproductive spore that is drawn into the needle via the same pathways, called *stomata*, that allow

Today
8,000 years ago

Atlas of Alberta

Rosalie LaRue/National Park Service

Left:
Whitebark pines (*Pinus albicaulis*) hug the tree line in Yellowstone National Park. These pillars of the ecosystem feed many wildlife species. Bushy needles moderate soil erosion by storing and slowly releasing the melting snow.

Below:
Blister rust has a complex life cycle. Spring rains grow telia stems on the underside of leaves in ribes shrubs. Bottom: rust-prone Indian paintbrush (*Castilleja coccinea*), the parasite's alternative host.

gases such as carbon dioxide, water vapors and oxygen to move rapidly in and out of the needles.

The following spring, pycnia form along the branch where the needles had been infected. The flask-shaped rust fungi reproduce asexually and create pycniospores over one or two years on the margins of the small canker on the branch in the spring. In the following year, aecium, a cup-like structure of rust fungus that contains chains of aeciospores, forms on the branch in the spring, and then flows with the wind across the mountain, hoping to land on a currant or gooseberry bush. Aecia attach to the leaves of the shrubs and cause yellow spots to form on top of them.

Within a few weeks, the presence of aeciospores on the leaves causes uredinia to hatch on the lower leaf surface, forming urediniospores, which fly away to infect other currant or gooseberry shrubs. Another rust fungus form, telia columns, begins to populate the leaf, and teliospores germinate to produce a four-celled creature called *promycelium*, on which four basidiospores develop. When the fall winds pick up, the basidiospores blow off the leaves of currant bushes and land on whitebark pine needles to begin the life cycle over again.

The impact of natural forest insects and diseases on coniferous tree species varies considerably, depending on the pest. The forests of the Northern Rockies contain well over 100 types of native insects and diseases, including sap-sucking insects, wood borers, defoliators, bark beetles, rusts, wilts, cankers, root diseases, and mistletoes. In actively

Donald Duerr/U.S. Forest Service

managed forests on private and state land in Idaho, foresters keep a close watch over insect and disease issues and often harvest diseased trees to avoid widespread epidemics. But timing is key. Problems need to be addressed swiftly to avoid bigger issues down the road.

Tom Eckberg, an entomologist for the Idaho Department of Lands, notes that ongoing forest management is the best way to prevent insect outbreaks, particularly the harmful bark beetle outbreaks. "It's a whole lot easier to prevent a beetle outbreak than cut yourself out of one," he says.

On Forest Service lands, where most of the whitebark pines reside, the agency hasn't traditionally tried to manage whitebark pine trees because they have no commercial value. The trees reside in remote high-elevation areas that rarely have any roads nearby, so active management is difficult if not impossible. But because of the ecological value of whitebark pine, and the landscape-scale losses of the trees from blister rust and mountain pine beetle, the agency's approach is changing. What can be done?

When *Cronartium ribicola* began to prey on white pine trees in the first half of the 20th century, it was considered a major crisis. White pine was the most prized merchantable tree species in Idaho and the Inland Northwest. The trees grew tall and straight, the wood was easily milled, and the loggers cut them down by the thousands, helping to settle northern Idaho towns like Coeur d'Alene and Potlatch. In 1910, there were 72 sawmills operating in three northern Idaho counties—Kootenai, Benewah and Shoshone, cranking out lumber created from the tall white pine trees. From 1925 to 1934, loggers harvested an average of 430 million board feet of white pine each year.

Left:
The cascading effect of the rust epidemic has sapped the biome's defense against other species of pests. Pictured: tree-boring Asian longhorn beetle, a recent invader from China.

Below:
Foresters struggle to understand why the exploding pine beetle infestation is more severe in forests blighted by rust. One explanation may be that both the beetle and the fungus prefer small-diameter trunks.

Ward Strong/B.C. Forest Ministry

Adults and larvae introduce mold into the tree's soft tissues.

Adult

Larva

Pupa

Sean Twiddy 2010

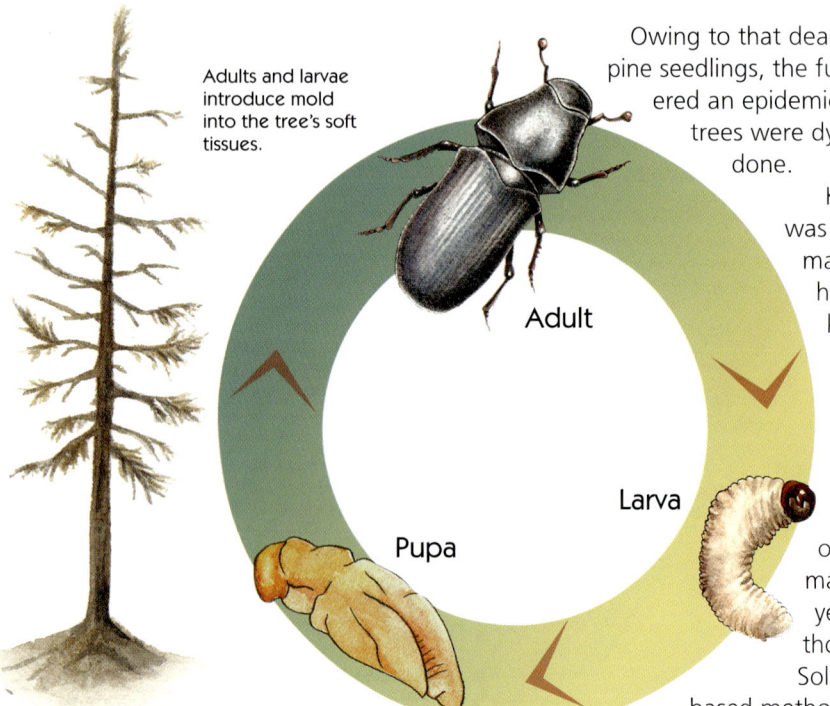

Owing to that deadly importation of rust-infected white pine seedlings, the fungus' impact on white pines was considered an epidemic by the 1940s. Thousands of white pine trees were dying on the stump. Something had to be done.

Knowing that the *Ribes* genus of shrubs was key to the blister rust's life cycle, the first major effort to save the white pine was to harvest the shrubs in great numbers and kill them with an herbicide, sodium chlorate (used today to bleach paper). Civilian Conservation Corps crews sprayed the stream banks of many streams and rivers with the herbicide, and they harvested many tons of shrubs and burned them. They even tried to inject antibiotics into the bark of white pine. Approximately $150 million was invested over 50 years to control blister rust, but none of those methods worked.

Solutions turned toward a more scientifically based method, trying to raise blister rust–resistant seeds in a laboratory setting. This effort started at the Forest Service science lab in Moscow, Idaho, in the late 1950s, where they successfully produced rust-resistant white pine seeds and trees. Since 1970, more than 200 million seeds have been produced at the University of Idaho orchard. Rust-resistant seedlings were grown at the Moscow nursery and inoculated with blister rust fungus to see how they survived. The results have been

Left:
Long harsh winters can kill the pine beetle larvae. Shorter winters and rising temperature have spread the beetle infestation north into Canada's lodgepole pines.

Below:
Mountain pine beetles bore through the bark. Like the blister rust, the beetles transmit a disease that constricts the tree's water supply. Whereas the rust leaps from branch to branch, killing in stages, beetles can kill a pine in 14 days.

mixed. The Forest Service's Coeur d'Alene nursery also has been working on raising rust-resistant white pine and whitebark pine.

Efforts to restore whitebark pine trees began soon after the year 2000. Numerous pilot projects have been implemented in the Northern Rockies, trying a combination of prescribed burns, planting rust-resistant trees, removing subalpine fir trees, which have been overcrowding healthy whitebark pine in some locations, and more. Forest Service and Park Service research officials have been tracking these efforts and recommending the best solutions on a landscape scale.

The Lowman Ranger District of the Boise National Forest got involved in the effort early on, as wildlife biologist Nadine Hergenrider spoke with silvaculturalist Pete Wier about the importance of the whitebark pine seeds to a multitude of bird and mammal species. The district did an aerial survey to document the status of whitebark pine at the time, and set up restoration projects on Whitehawk Mountain in Bear Valley and Scott Mountain near Garden Valley. The district completed an environmental analysis required by law and went to work.

One of the first orders of business was to remove subalpine fir trees that were overcrowding whitebark pine. This has been occurring as a result of the historical absence of wildfire. At Whitehawk, they physically removed subalpine fir and sold the trees to a company that hauled the low-value species to a sawmill. On Scott Mountain, they either burned the alpine fir trees or "girdled" them, meaning they cut a ring through the bark and cambium layers to kill the trees.

Oakland Press

Left:
Prescribed burns replicate nature's way of thinning and cleansing the forest, replenishing its cycle of life. Whitebark pines thrive in fire-blackened landscapes. Fire suppression has accelerated the whitebark's demise.

Below:
Cronartium ribicola can take years to blister a pine with cankers. Even after 1910, when blister rust reached Seattle, tree importers and nurseries unwittingly spread the disease. By 1940 the rust had jumped from Idaho into Montana.

The prescribed burns, which always are timed in the spring or fall so as not to start a large wildfire, not only got rid of the alpine fir but also left small patches of blackened soil on the forest floor. The foresters understood that Clark's nutcrackers prefer the blackened soil for caching seed. Several years later, new whitebark pine seedlings are growing in those blackened patches, the product of human-enhanced natural regeneration.

In other burned patches, the Lowman District hand-planted whitebark pine seedlings to see how they would succeed. As we toured the site, we saw a number of seedlings that looked healthy and seemed to be growing vigorously. We also saw a number of healthy whitebark pine trees that seemed small but might be as old as 40 or 50 years. Those trees will be old enough to bear cones soon.

The seeds for the seedlings were collected from cones on healthy whitebark pine trees on Lowman's Scott Mountain. People like Lazarus have been

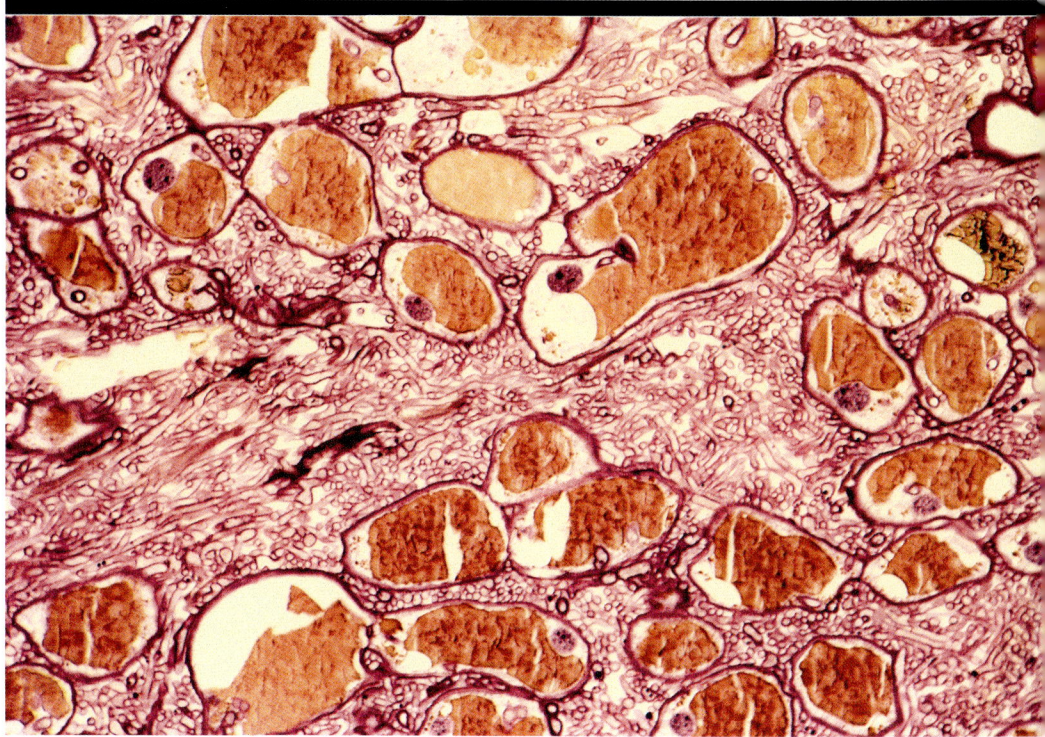

Science Source

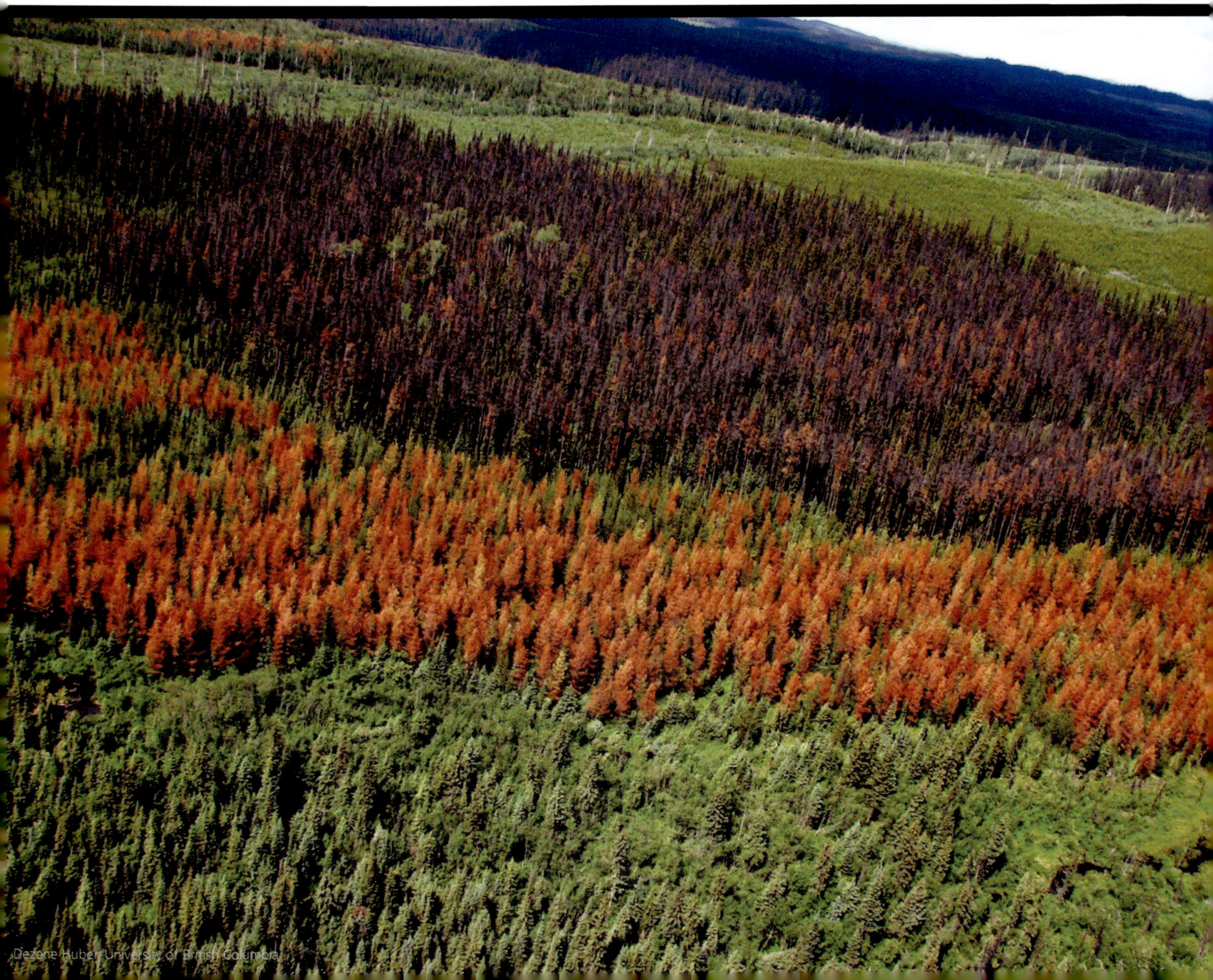

Dezene Huber/University of British Columbia

Left:
Orange and purple-red swaths of trees dead and dying may be the dire future of an ecosystem on the brink. Already in the icy highlands, more whitebarks are dead than alive.

Below:
Congress, in 1912, first responded to the epidemic with a quarantine of currant and gooseberry ribes. New Dealers reignited but ultimately lost the war to eradicate ribes. Pictured: gooseberries (*Ribes uva-crispa*).

busy identifying and marking these healthy trees for a number of years, making note of their GPS coordinates for future reference. Foresters climb the trees to place a special netting over the cones to prevent Clark's nutcrackers, red squirrels, bears and other birds from harvesting them. When the cones are ready for harvest, foresters climb the trees, collect the cones and send them to the Forest Service's Lucky Peak nursery. The nursery is growing about 12,000 whitebark pine seedlings for eventual outplanting in the Boise National Forest in the next couple of years.

The Forest Service's Coeur d'Alene nursery also collects cones from healthy whitebark pine trees in numerous national forests in northern Idaho, Montana, Wyoming, Washington and Oregon and produces seedlings for outplanting at a much larger scale. The nursery's location in Coeur d'Alene is wet and moist through much of the year, and it's considered ideal for growing rust-resistant seedlings for many national forests, nursery officials say. Over the past decade, whitebark pine cone collections from forests throughout the region have allowed the Coeur d'Alene nursery to ramp up the production of seedlings from 5,000 in 2008, to 95,000 in 2011, to 240,000 in 2013. "It's been one of our main focuses in the last few years," says Nathan Robertson, a horticulturalist for the Coeur d'Alene nursery.

All of those seedlings will be planted in national forests in the Greater Yellowstone Ecosystem in Wyoming, Montana and Idaho; the Sawtooth National Forest in Idaho; and national forests in Washington and Oregon.

Robert Keane, a research ecologist with the Missoula Fire Sciences Laboratory, is pleased to see tree-seedling production efforts ramping up as he

Below:
Blights, smuts, rots, and mildews top the Idaho Department of Agriculture's watch list of invasive pathogens. Pictured: Larvae of the invasive sawfly (*Pristiphora geniculata*) feed on leaves of the indigenous American mountain ash.

Wikipedia commons

recommends moving forward as rapidly as possible with more cone and seed-collection efforts and tree-planting at a broad scale to hasten restoration efforts for whitebark pine. Keane has been tracking the success and failure of whitebark pine restoration projects for a decade. In 2012 he authored a paper titled "A Range-

Left:
More than 7,800 species of fungi are commonly referred to as rust. Most stunt, weaken, and wither but seldom kill their host. Pictured: orange rust (*Arthuriomyces peckianus*) on a black raspberry leaf.

Below:
Cronartium ribicola reproduces with five different types of spores. Lighter spores can be transported by wind or insects. Teliospores with thicker cell walls endure winter at rest on the primary host.

Wide Restoration Strategy for Whitebark Pine (*Pinus albicaulis*)," which provides a detailed menu of options for preserving the species and the ecosystem.

"I'm totally positive that we can restore whitebark pine," Keane says. "The question for us as a society is whether we have the willpower to do it."

Part of the issue is funding, he notes. Less than $1 million a year is being spent nationwide to save whitebark pine, but about $3 million to $4 million is needed for restoration activities on an ongoing basis, Keane says. Areas that have been ravaged by the mountain pine beetle and white pine blister rust should be the highest priorities, he says.

Normally, in official wilderness areas such as "the Frank" in central Idaho, tree-planting is not allowed except in the case of emergencies, when natural regeneration will not occur quickly enough to save a species. National Park Service officials have decided that planting whitebark pine in parks and wilderness areas is necessary, Keane says. "Glacier National Park has a very active cone- and seed-collecting program right now, and they're constantly out there planting new seedlings," he says. "They've jumped on the program full-bore."

But the Forest Service has been wringing its hands about whether to allow rust-resistant trees to be planted in official wilderness areas. The concern is if rust-resistant seedlings carry any genetic material that is not native to the wilderness, perhaps that could create more problems down the road, Keane says.

Larry Eifert

Left:
Whitebark pines, imperiled in the highest places, have come to symbolize the fight to protect the vanishing wild. Canada listed whitebarks as an endangered species in 2010.

Below:
A fungal pathogen called *Gymnosporangium juniperi-virginianae*, or cedar rust, marks apples with glistening lesions. Spores spew from its jellylike horns.

 In the meantime, the Forest Service's geneticists have been conducting experiments in the laboratory, working on raising whitebark seedlings with rust-resistant seeds and then testing their ability to survive more attacks by white pine blister rust by inoculating the trees with the pathogen and seeing how they do. Special rust-resistant "orchards" are being set up in different climatic regions of the North Rockies to carry on experiments as another way to preserve the species.

 As Lazarus and Swenson wrap up the field trip on Scott Mountain, Swenson takes on a decidedly upbeat tone after we've seen more of the newly planted seedlings growing out of the blackened soil. Projects like the one on Scott Mountain may be part of what defines the Forest Service's future direction. Restoration is the new buzzword that applies to many activities leading to a healthier forest. Perhaps a growing effort to restore whitebark pine could become a building block for the agency's future.

 "One day, this might look like it did before European settlement, with big open fields of some big ol' whitebarks growing everywhere," Swenson says, grinning at the thought. "This is looking awesome! We have multiple generations of trees growing here, and diversity in age classes. See, something worked!"

Matthew Wills

Alien Invaders

Non-native invaders from our globe's every corner cause billions of dollars in damage in the Pacific Northwest. Some are harmless, but many kill native species, altering ecosystems. Rust from Asia blisters the pines. Feral pigs, migrants from Spain, transmit flu and tuberculosis. Mussels from the Caspian Sea, having colonized the Great Lakes and spread to the Colorado River, are poised to jump the Idaho Tetons into the Snake River canyonlands.

Since 1900, with the passage of the Lacey Act, Congress has targeted invasive pests as threats to agriculture. The U.S. Geological Survey names more than 6,500 species of invaders. Damage estimates exceed, annually, $130 billion. Invasive insects top $13 billion in damage to American crops.

bugvibes.com

Nutria *Entamoeba histolytica* **Monk parakeet**

Pictured: Japanese beetles infest Idaho's Boise Valley. Insets: Nutria "river rat" from Brazil, a threat to American wetlands; Europe's *Entamoeba histolytica*, a source of dysentery; monk parakeets from Argentina, a rival to native songbirds.

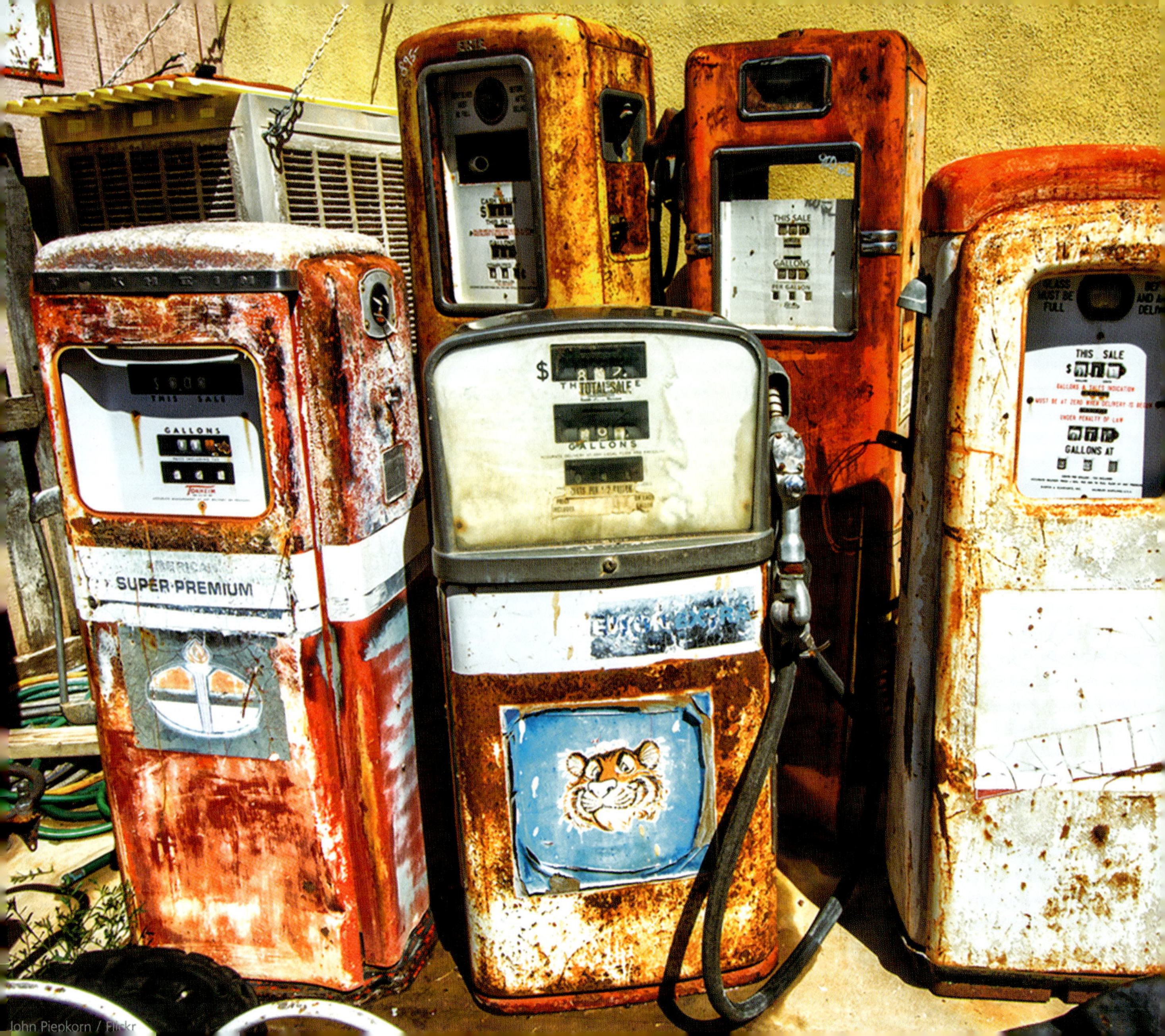

John Piepkorn / Flickr

guzzling crude

harnessing the power of bioremediation

7

Science Source

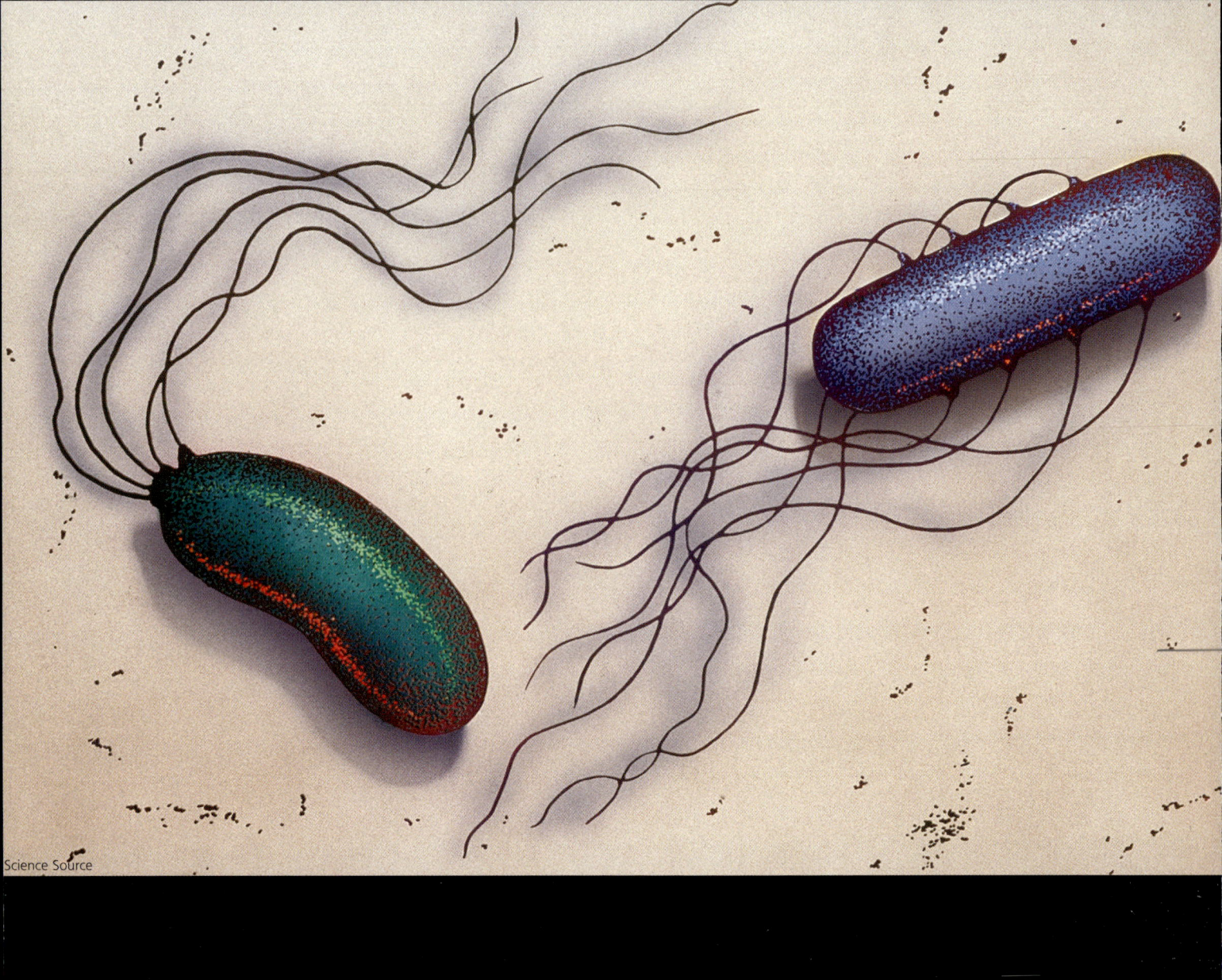

Left:
To *remediate* is to fix a problem. To *bioremediate* is to cleanse water and land with a healing symbiosis of microscopic life. Pictured: *Bacillus* (bottom left) and *Pseudomonas* bacteria, both bioremediators.

Below:
New Deal photographer Russell Lee celebrated the wholesome purity of water in the Boise Valley. Previous: Rusted pumps mark the corrosion of buried tanks and piping that threaten our water supply.

It's a bright sunny day in early May at the Lakeshore Market, located south of Nampa just off of Idaho State Highway 45. It's a convenient location where people can fill up their vehicles with gas and buy supplies before they head south to the Snake River or the Owyhee Mountains. In 1989, a customer was filling a cup with ice before filling it with soda pop inside the market, and he thought he smelled an odor of gasoline in the ice cubes. The customer notified the Idaho Department of Environmental Quality (DEQ), and DEQ responded by testing the water. It turned out that the market's water well had been contaminated with petroleum fuel that had leaked from the underground storage tank system at the gas station.

"Several businesses located behind the gas station said they had fuel in their water, too," says Eric Traynor, Brownfields Response Program manager for DEQ. "The previous owner pulled the tanks and installed a new, deeper well to serve everyone here."

The petroleum leak at Lakeshore Market was, unfortunately, one of many discovered in the Treasure Valley at the time. Mark Van Kleek, who used to work in the DEQ regional office in Boise, recalls that he was really busy in the late 1980s and early 1990s, responding to fuel spills and leaking underground storage tanks almost on a daily basis. "I remember that we had 65 sites that we were investigating at the same time," Van Kleek says. "You'd hire a contractor to go out to a site to dig up the tanks, and as soon as they started digging, you knew right away—you could see it, you could smell it, it was obvious. A lot of those tanks needed to come out of the ground and get

Library of Congress

replaced, but most people didn't have the money to fix them."

Most of the underground fuel tanks and lines were made out of bare steel, and over time, they would corrode and begin to leak. Small releases during fuel deliveries were another common occurrence. A lot of those systems were used at dozens of "Ma and Pa"–type gas stations in Idaho and elsewhere throughout the United States. At the same time, the Environmental Protection Agency (EPA) issued new rules and regulations regarding underground fuel-storage tanks. The new rules required the installation of double-walled tanks, leak detection systems for the tanks and piping, spill buckets to catch any small releases, and standards for underground storage tank system operation and maintenance.

The new regulations seemed burdensome to service station owners at the time, but they were needed, as a multiplicity of fuel leaks had contaminated portions of shallow groundwater aquifers in Nampa, Boise and elsewhere in Idaho. People who had shallow drinking water wells located nearby were affected. "I remember my parents' well was impacted by a fuel spill out by the mall," says Van Kleek, a Boise native. "They had to hook up to public water, and the party liable for the contamination had to pay for the hookup."

But how does the groundwater get cleaned up? Even some 25 years later, the Lakeshore Market site still has substan-

Left:
Sunflowers metabolize petroleum hydrocarbons. Where bacteria work underground to degrade petro pollutants, sunflowers neutralize the toxic contamination of oil-soaked surface soils.

Below:
A welder uses a cutting torch to dismantle a leaky gas tank. Even a pin-prick leak of one drop per second releases 400 gallons each year. Inset: vintage sign from the era of the motor tourist.

tial levels of petroleum products in the soil and groundwater underneath the service station. In May 2014, DEQ officials and a professional cleanup contractor were on-site to conduct a number of groundwater tests and check on the levels of oxygen, carbon dioxide, methane and volatile organic compounds.

Bruce Wicherski, the Voluntary Cleanup Program manager for DEQ, opened a groundwater monitoring well at the Lakeshore Market that was downgradient of the fuel spill and took a reading. The carbon dioxide levels in the subsurface vapor were 2.8 percent, compared with typical ambient air levels of approximately 0.3 percent. That meant the ozone-sparging system they had installed on-site was super-charging the groundwater and underground soil profile with oxygen, which, in turn, stimulated the activity of naturally occurring carbon-eating microbes in the subsurface soil.

"They're working," Wicherski says, referring to the microbes. "If there's lots of carbon dioxide, there's lots of bugs operating. You would expect that."

Cleaning up groundwater after fuel spills have occurred can be expensive and time-consuming. The technology for cleaning up spills is getting better and more advanced as time goes on. DEQ officials point out that *bioremediation* or *biodegradation* of fuel spills—that is, stimulating naturally occurring microorganisms by injecting ozone, oxygen or fertilizers into the ground—is typically considered a secondary treatment.

Liren Chen

Jack Rendulich / *News-Tribune & Herald*

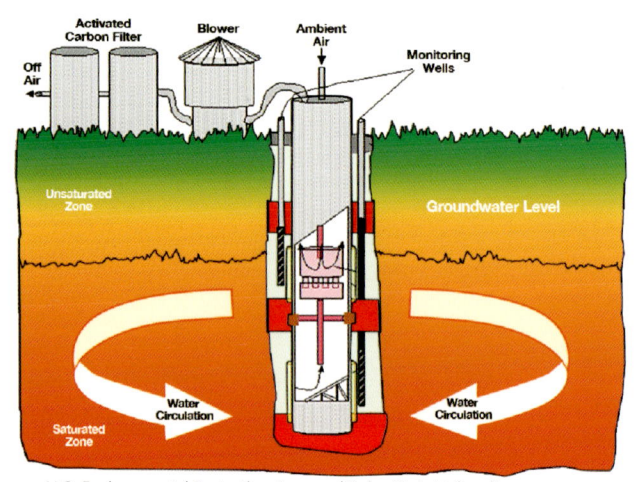

Activated Carbon Filter · Blower · Ambient Air · Monitoring Wells · Off Air · Unsaturated Zone · Groundwater Level · Saturated Zone · Water Circulation

U.S. Environmental Agency / Boise State University

The primary ways of treating acute underground fuel spills, DEQ officials say, are to use ozone-sparging, soil vapor extraction, or pump-and-treat, in which water is pumped out of the ground and run through a treatment process to extract the petroleum products, and then those volatile organic compounds are burned off. In a typical ozone-sparging/soil vapor extraction process, oxygen is pumped into the groundwater via a sparge well to strip dissolved petroleum compounds from the water, while petroleum vapors are extracted via an extraction well placed above the water table. The vapors are then vented to the atmosphere through a stack.

"The vapors are tested during a pilot test to determine if they exceed permitted levels," Traynor says. "If vapor concentrations exceed permitted levels, then they are treated to reduce concentrations prior to discharging to the atmosphere. Treatment of vapors is usually done using either a catalytic or thermal oxidizer."

In the case of the Lakeshore Market situation, they're

Oil tankers and tanks once lined the rail yards of Union Pacific. In Pocatello, near the city's intake wells, the UP evacuated 14,000 tons of oily sludge seeping through the water table. Diagram: Hydrocarbons converted to vapor bubble up to the surface through an elaborate system of air-injection wells.

CPDX 107116
CAPY 35700 US GAL
CAPY 28062 IMP GAL
CAPY 127 568 L

using ozone-sparging to break down the petroleum products in the groundwater, and by adding additional oxygen, they also stimulate the microbial activity. This system is what the owner can afford. Hardeep Singh bought the service station in 2006 without realizing that the soil and groundwater were contaminated under the service station. "I got caught in the middle of it," Singh says. "I want this site cleaned up pretty badly. I want to get it cleaned up as soon as possible and be done with it."

Because he is more than two property owners removed since the contamination was discovered in 1989, Singh qualified for the DEQ brownfields program, which is funded through the EPA. This is a national program in which financial assistance is provided to allow environmentally contaminated sites to become more economically productive in the long term. In 2008, the Idaho Legislature passed the Community Reinvestment Pilot Initiative, providing $1.5 million to DEQ to fund 10 remediation projects around the state. Eligible participants receive a rebate of 70 percent of eligible cleanup expenses up to $150,000 once a cleanup project has been completed. Lakeshore Market is one of the 10 pilot sites.

Bruce McAllister, National Archives

Left:
Petroleum's stress on ecosystems can be hard to detect. In 1977, a modest spill in the Baltic Sea killed twice as many ducks as the massive 1989 slick in Alaska's Prince William Sound.

Below:
Strains of *Pseudomonas fluorescens* have been genetically engineered to eat copper, neutralize TNT and detoxify cyanide. Rigid cell membranes help the microbe survive.

Singh also received a low-interest loan through the brownfields revolving loan fund to finance the cleanup activities. As a result, Singh is qualified to receive a $150,000 rebate when the cleanup is complete. This will be applied to his loan, DEQ officials say. To receive assistance, Lakeshore Market had to qualify among a state-ranking system of brownfield sites.

One reason it's taken so long to clean up the site is that the groundwater in the area is extensive, lying on top of hard basalt rock, Traynor says. "The groundwater out here is like a lake; it has a very flat gradient," he says. "The contamination is fairly deep. With sites like this, with fractured basalt, it can be difficult to determine where the contamination is going as it can hang up in the fractures and voids in the basalt, making it difficult to remediate. On sites where the groundwater contamination is in sands and gravels, they clean up faster because site conditions are more favorable."

The ozone-sparging system now working at the Lakeshore Market was used to clean up a fuel spill site in Homedale. "This system worked well," Traynor says. "The system was running for seven months when groundwater monitoring results indicated that the groundwater remediation goals had been achieved. That was a fairly short remediation event, compared to other sites that have run ozone-sparging and soil vapor extraction systems for two to three years or more."

The ozone-sparging system injects ozone into a number of groundwater wells in the contaminated area. The ozone breaks down the petroleum products underground through chemical oxidation and enhances aerobic bioreme-

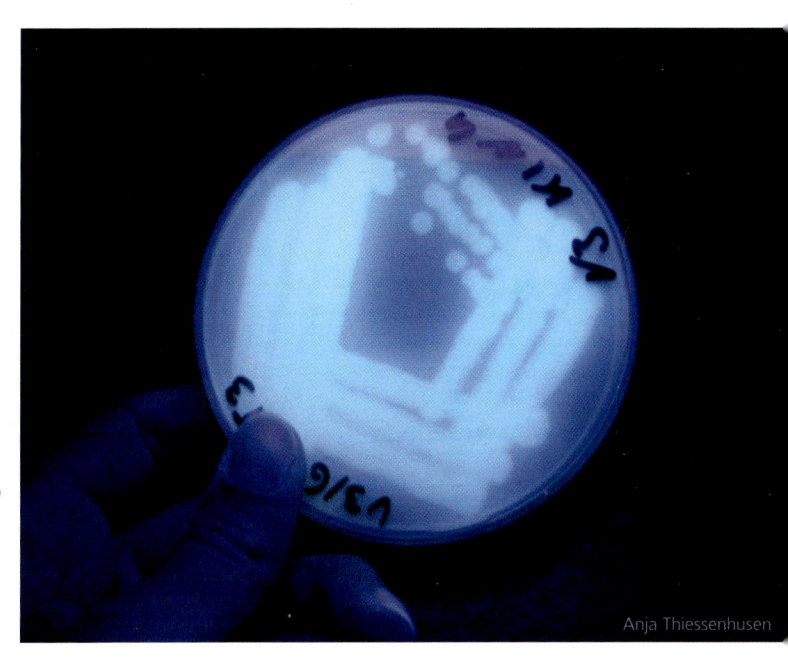

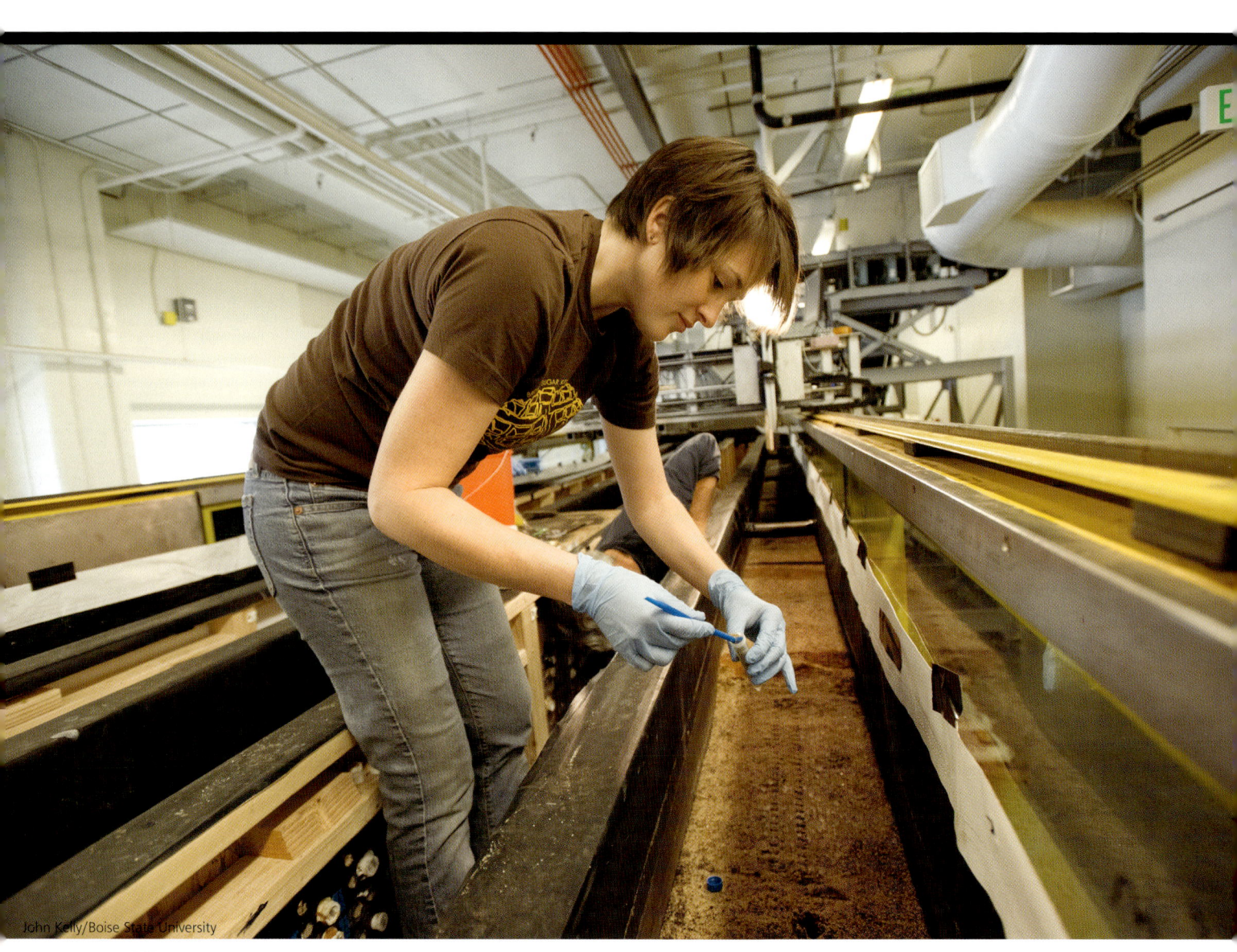

John Kelly/Boise State University

Left:
Boise State's Tiffany Farrell, a graduate student in hydrologic sciences, samples soil from a laboratory flume at the Idaho Water Center. Farrell's research concerns bacteria that reduce emissions of nitrous oxide.

Below:
Pseudomonas is a versatile genus with more than 100 bacterial species. A beast and a benefactor, the genus can metabolize industrial toxins and spread superbug staph infections.

diation. An EPA website explains how it works: "Ozone is 10 times more soluble in water than pure oxygen. Consequently, groundwater becomes increasingly saturated with dissolved oxygen as the unstable ozone molecule decomposes into oxygen molecules. About one-half of dissolved ozone introduced into the subsurface degrades to oxygen within approximately 20 minutes. The dissolved oxygen can then be used by indigenous aerobic hydrocarbon-degrading bacteria."

So it's a two-part cleanup process: (1) breaking down the petroleum products with ozone and chemical oxidation, and (2) stimulating the microbial community to consume the fuel.

Traynor notes that ozone works well to break down the petroleum products in situ in the ground. There is no waste product. "It's like hitting it with a sledgehammer," he says.

Adds Wicherski, "With oxygen present, the bugs get happy and start multiplying and eating."

I asked the DEQ officials at the cleanup site if they knew, specifically, what natural microbes in the soil were consuming the petroleum products. They looked at me and shook their heads. "We don't. We just want to know that they're working," Wicherski says.

Kevin Feris, a professor of microbiology at Boise State University, knows something about the microbes that consume fuel spills. He did some postgraduate work on the topic at the University of California, Davis. As a microbiology expert, he also knows how to culture microbes from soil or groundwater. So Feris helped us identify a few representative microorganisms working on the cleanup project at Lakeshore Market. Under the supervision

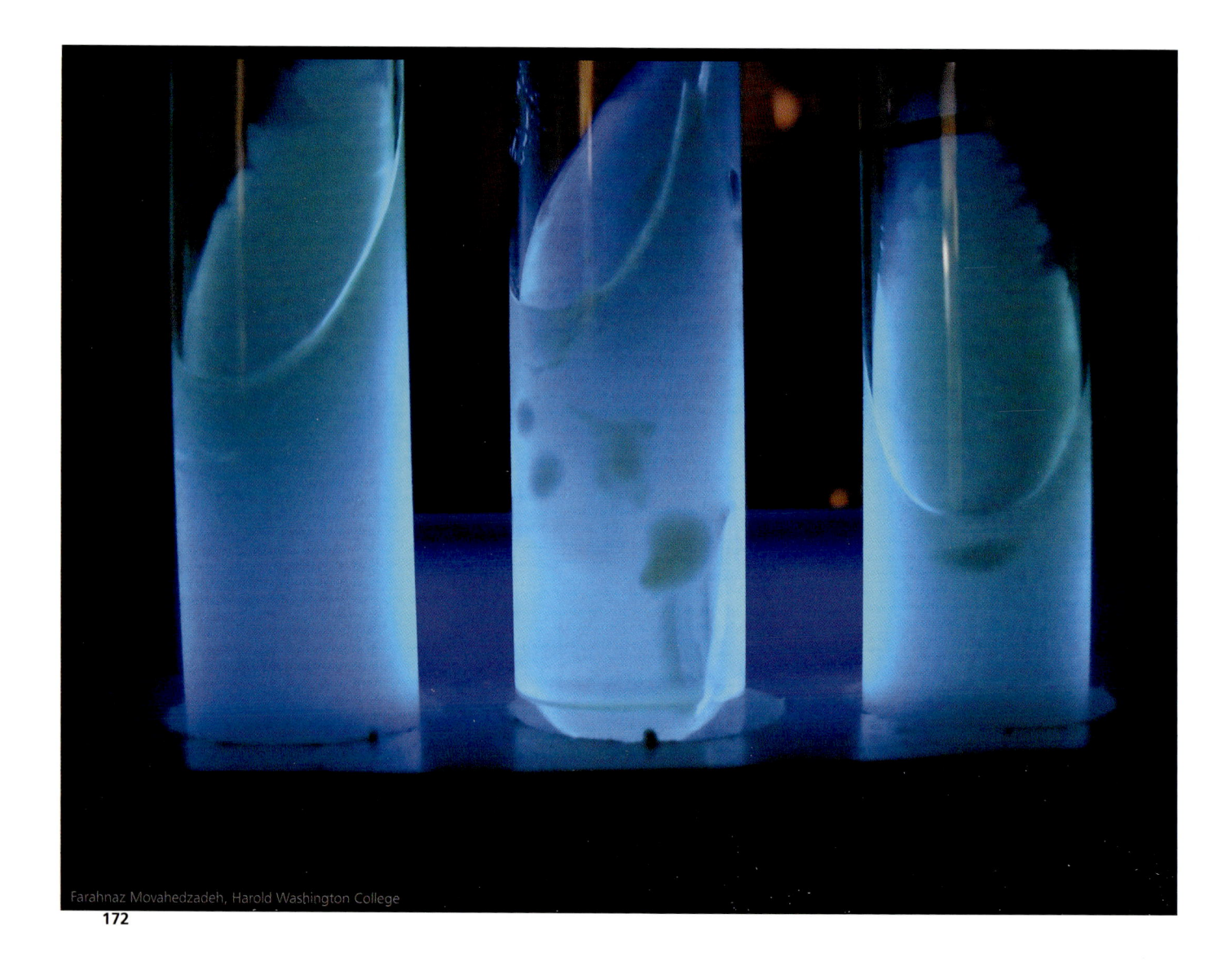

Farahnaz Movahedzadeh, Harold Washington College

Left:
Toxic water from the Chicago River is helping researchers understand how *Pseudomonas fluorescens* degrades human sewage. Ultraviolet light shows the microbe's blue-green fluorescence.

Below:
Pipettes, pumps and syringes are common tools for liquid extraction. Pictured: a pipette, used to sample a measured volume of liquid. Inset: Kevin Feris works with a student in his microbiology lab.

of DEQ's cleanup contractor, he drew groundwater samples from upgradient and downgradient wells at the market.

In the downgradient well, Feris says, "Whew, the water really smelled like a can of gas. It looked really black."

To isolate the microbes, Feris explains that he can "create an environment in which they can grow" in round petri dishes in the biology lab. The exercise is similar to what his students are required to do in his General Microbiology 303 class. The petri dishes are sterilized and coated with a clear substance called agar, a common substrate on which the microorganisms can grow. Feris lines up the petri dishes on the gray lab counter and sets up a Bunsen burner and flame to sterilize a sharp, metal-tipped instrument called an inoculating needle. "The bacteria tend to form different colonies with different shapes and colors," he says. "Each colony is a population of cells that come from one or two individuals. As the cells divide and multiply, the colonies form on the agar surface."

With the fine-tipped instrument, he introduces the bacteria from the upgradient and downgradient wells into the petri dishes. How many cells might be picked up by the needle each time? "I probably just picked up a few thousand cells" with that one dose, he says. He puts a lid on the petri dishes and then places them into an incubator at a similar temperature to the underground environment, about 20 degrees Celsius (68 degrees Fahrenheit). He lets the cultures grow for a couple of days.

After that, he pulls the petri dishes out of the incubator. Indeed, tiny colonies are beginning to form in the dishes. Very small colonies—a series of tiny dots—were growing in the samples from the upgradient well, where there is less petroleum contamination, and many more colonies were growing in the samples from the downgradient well with more contamination. "We have some nice individual isolated colonies," he says.

Putting the upgradient and downgradient petri dishes on the lab counter, Feris translates using an analogy of looking at lights below when flying over the United States. "With these (upgradient) samples, we're flying over Kansas,

Left:
Scientists have classified more than 200 microbial species that can break down hydrocarbons. Pictured: A flotilla of *Oleispira antarctica* (circled) swims toward a droplet of oil.

Below:
Whiplike flagella help *Pseudomonas fluorescens* cope with extremes. Mobile and quick to mutate, the bacterium can endure exposures to radiation that would kill most any other creature on Earth.

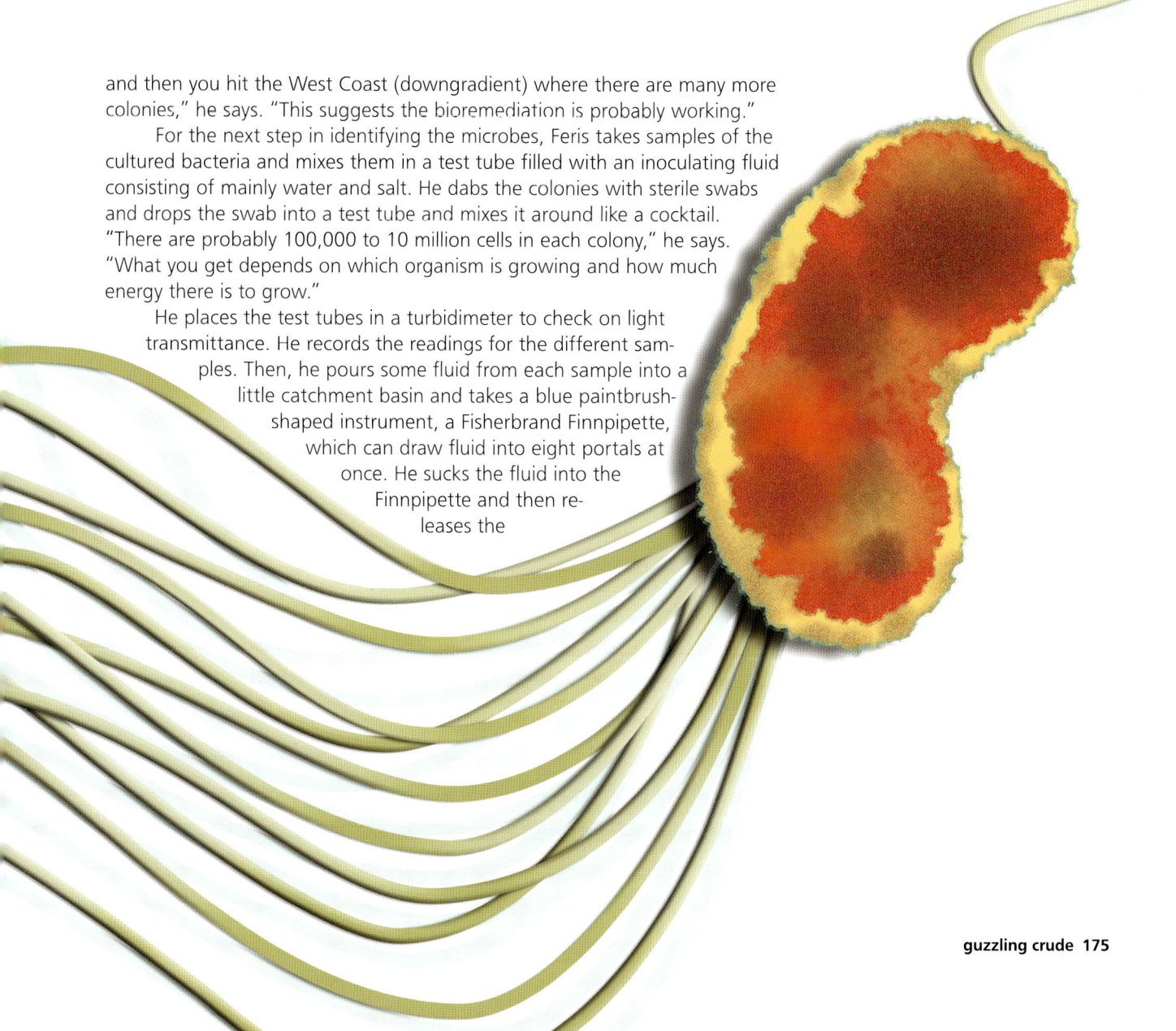

and then you hit the West Coast (downgradient) where there are many more colonies," he says. "This suggests the bioremediation is probably working."

For the next step in identifying the microbes, Feris takes samples of the cultured bacteria and mixes them in a test tube filled with an inoculating fluid consisting of mainly water and salt. He dabs the colonies with sterile swabs and drops the swab into a test tube and mixes it around like a cocktail. "There are probably 100,000 to 10 million cells in each colony," he says. "What you get depends on which organism is growing and how much energy there is to grow."

He places the test tubes in a turbidimeter to check on light transmittance. He records the readings for the different samples. Then, he pours some fluid from each sample into a little catchment basin and takes a blue paintbrush-shaped instrument, a Fisherbrand Finnpipette, which can draw fluid into eight portals at once. He sucks the fluid into the Finnpipette and then releases the water samples into a Biolog microplate. The microplate tubes already are populated with different media, including carbon. After a 24-hour period of incubation, the tubes on the microplate turn color, indicating a positive result. The colors help Feris identify the species of microbes present in the sample. Biolog provides a software package with a color key for identifying species, Feris explains. "Using this approach, you can quickly key out 15,000 to 30,000 different species. It's a big time-saver."

Even before he gets the results, Feris postulates that one of the microbes will be a *Pseudomonas* genus. "They're a well-described genus that does a good job of breaking down hydrocarbons and are commonly observed in contaminated groundwater when oxygen is present," he says. "Some of the small round colonies, and the slimy looking ones, they look like *Pseudomonas*."

The following week, Feris writes me to indicate that he's isolated *Pseudomonas putida* and *Pseudomonas fluorescens*. "This is kind of what you'd expect," Feris says. "*P. fluorescens* is a common soil bug that has some BTEX degrading capabilities, and *P. putida* is the type of species commonly found in aerobic subsurface BTEX plumes. Pretty cool."

©Laura Gilmore

Procter & Gamble

Left:
Brownfield blight within the City of Boise includes former laundry and dry cleaning sites. Pictured: washer woman at an abandoned Boise laundry. Inset: Tide detergent enriched with sodium lauryl sulfate, a carcinogen.

Below:
Some bacteria immobilize groundwater pollutants by keeping them solid and restricting their flow. Pictured: Sulfate-reducing *Desulfovibrio desulfuricans* surrounds crystals of lead.

BTEX is an acronym describing the mix of chemical constituents commonly found in petroleum plumes—benzene, toluene, ethylbenzene and xylene. The DEQ monitoring report from Lakeshore Market showed high levels of benzene, 1,2-dichloroethane, ethylbenzene, methylene chloride, naphthalene, 1,2,4-trimethylbenzene and 1,3,5-trimethylbenzene in the groundwater. The levels of those constituents exceeded state standards in three of the on-site monitoring wells and one off-site well downgradient of the service station.

Pseudomonas putida, it turns out, is a bit of a dynamo when it comes to bioremediation. A species of bacteria, *P. putida* has been isolated at numerous bioremediation sites in North America. It guzzles hydrocarbons several times faster than other *Pseudomonas* species, according to research work by Ananda Mohan Chakrabarty, an Indian-American microbiologist. Chakrabarty found a way to genetically modify *P. putida* to make it work even faster on bioremediation sites.

Chakrabarty filed for and received a patent for his genetic modification work with *P. putida*. It's one of the few patents that have been issued for a live organism. The patenting case was fought in the U.S. Supreme Court, but Chakrabarty won and got his patent.

Pseudomonas putida is a rod-shaped single-celled bacterium, hailing from the phylum Proteobacteria, named for the Greek god of the sea,

Proteus, because of the diversity of species in the group. The bacterium is found in most soil and water environments where oxygen is present. It not only degrades organic solvents like toluene but also converts styrene oil to biodegradable plastic. It has flagella for mobility. Under a microscope, it looks like an oblong blob with several strands of long flagella in the rear. *P. putida* apparently has the ability to sense when new sources of protein and oxygen are available, and on receiving that signal, it eagerly seeks out the new source of food. It reproduces asexually through binary cell division. It does not produce any spores.

Pseudomonas fluorescens is a well-known bacterium that assists in protecting some plants from parasitic fungi and phytophagous nematodes. In fact, it's being used as an experimental biocontrol agent on cheatgrass at Deer Flat National Wildlife Refuge in Nampa. *P. fluorescens* has bioremediation properties as well. It also comes from the phylum Proteobacteria. It has strong metabolic properties, and it has lots of flagella for mobility. It reproduces asexually.

Both *P. putida* and *P. fluorescens* have a rigid cell wall that allows them to survive under harsh conditions, such as living amid toxic hydrocarbons in the soil and/or water.

Feris noted, as he worked through the samples, that the bacteria he cultured would represent only about 1 percent of the species likely to be working in the plumes. But it gives us an idea of what species exist and how they work.

Left:
Bacteria thrive near the leaky tanks of the Hanford Nuclear Site. The iron-rich groundwater below the reactors helps the bacteria grow.

Below:
A worker bags a duck from the gluey tar of the 2010 Deepwater Horizon disaster. The spill exploded the beach population of bacteria that consume oil for fuel.

Experts in microbiology say that much work remains to be done to isolate more microbes that assist in cleaning up toxic oil spills underground or at sea. Only a small percentage of the species have been identified, but when researchers delve into the topic of how microbes clean up acute oil spills at sea, they are amazed at how fast some of the species work, and they've discovered new species.

After the Deepwater Horizon oil spill (also referred to as the BP oil spill), for example, a team of researchers from the Lawrence Berkeley National Laboratory in California collected 200 water samples from 17 deepwater locations in the Gulf of Mexico to check on microbial activity between May and June 2010. They found microbial cell concentrations had doubled to 5,510 cells/milliliter, compared to normal levels of 2,730 cells/mL outside the plume. They found a wide array of microbe species, including known petroleum-eaters such as *Oleispira antarctica*, *Oceaniserpentilla haliotis* and *Thalassolituus oleivorans*, according to *Science* magazine.

Their findings show the amazing natural healing powers of the oceans—and the Earth, says Terry Hazen, director of the ecology department at the Lawrence Berkeley National Laboratory. "Bioremediation holds great promise for some of our worst problems," Hazen told *Science*. "There is no compound, man-made or natural, that microorganisms cannot degrade."

For the best results, it takes a multitude of microbes to clean up a site because each species typically consumes one particular compound

Wikipedia

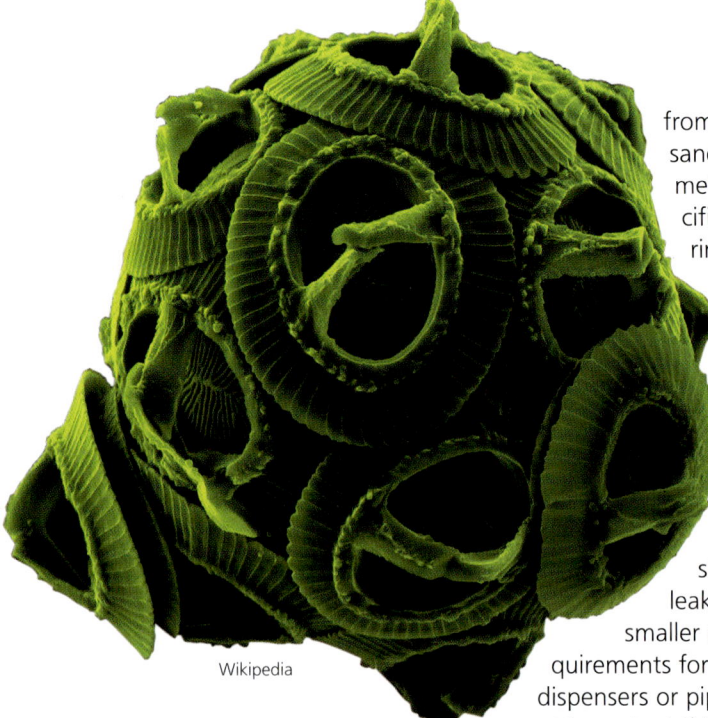

from a fuel spill, such as benzene. Crude oil contains thousands of different hydrocarbons, including gases like methane, and liquids, such as toluene. "They tackle a specific compound, and only that one," Andreas Teske, a marine microbiologist at the University of North Carolina at Chapel Hill, said in *onEarth* magazine. "In order to degrade a complete oil spill, you need a big community of specialized bacteria that act in concert."

As Idaho DEQ officials caution, microbes do not accomplish fast work overnight, but they do heal things over time.

On a statewide basis, the number of fuel spills in Idaho is decreasing, DEQ officials say. The new underground storage tank regulations were a key part in slowing the trend. "Tanks are not the main source of leaks anymore," Traynor says. "The releases are generally smaller as they are detected quicker due to leak detection requirements for tanks and piping, and they're mostly associated with dispensers or piping. There can be a lot of different factors. One of the most important things is to stay on top of your fuel inventory, system maintenance, and make sure the equipment is in good working order. Keep good records."

"The trend is down," Wicherski adds. "We go by events, and we currently have 117 events or active remediation sites out of 1,450 total sites. Sixty-five of the active sites are in the southwest Idaho region. Ninety percent of all the sites are closed."

The number of underground storage tanks currently registered in the state also has dropped, from 12,000 in 1995 to 3,500 in 2014. That indicates

Left:
Photosynthetic algae may be the bright green future of sewage remediation. Strains grown in the lab remove nutrients from wastewater. The organic waste can then be converted to liquid biofuel.

Below:
When big fish eat smaller fish, pollutants are concentrated through the process scientists call biomagnification. Bacteria engineered to metabolize toxins may offer the best defense.

that the number of service stations operating today has decreased, and the ones that do operate must be in compliance with all of the rules. Judging from the growth of food marts and service stations in the Boise Valley, running a mini-mart can be a good business.

At Lakeshore Market, Hardeep Singh may have inherited a toxic mess, but at least he bought a station that has modern double-walled steel tanks for diesel, premium and unleaded gas with fiberglass-reinforced plastic piping and an automatic tank-gauging system for leak detection. As the cleanup occurs with the assistance of the DEQ's brownfields revolving loan program, he may be able to expand his business in the southern outskirts of Nampa someday. He and his wife also own a Chevron service station in Boise at the corner of Franklin and Curtis.

"We're hoping for a bright future," Singh says. "We like living here."

Sportsman pages

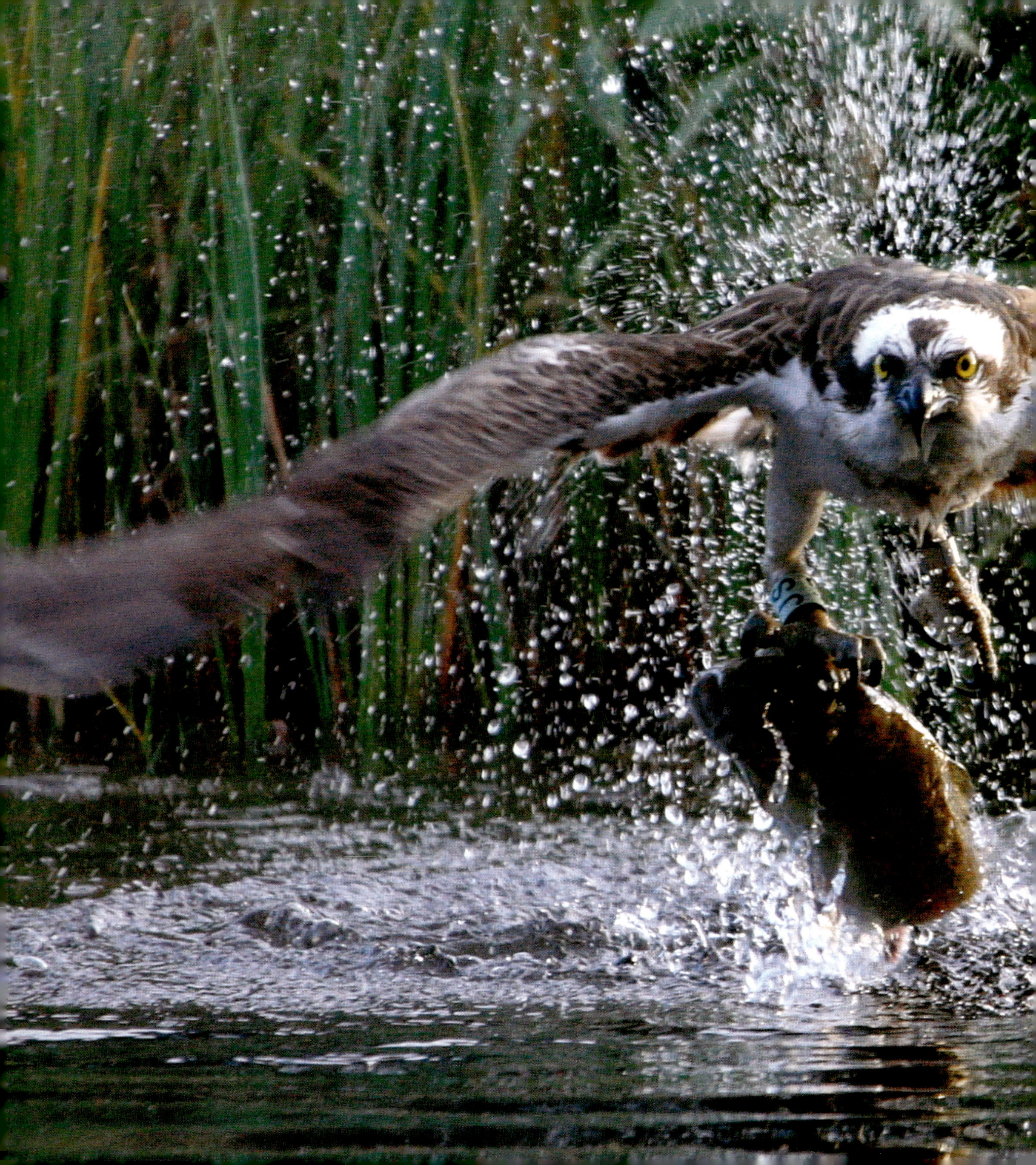

Climbing the Toxic Food Chain

An osprey's feathers tell tales of the poisoning of an ecosystem. In North Idaho's Silver Valley, where heavy metals from silver mining drain to the bottom of Lake Coeur d'Alene, ospreys dive for the trout that jump for the flies that feast on the toxic plankton. The toxicity magnifies with every link up the aquatic food chain. Biomagnified, the toxins kill fish and vegetation. Osprey feathers show mercury concentrations up to 60 times the EPA safety standard.

By 1983, when Silver Valley became an EPA Superfund site, one in four of the district's toddlers had bloodstreams toxic enough to damage the kidneys and brain. Pictured: A leg ring helps scientists track osprey migration. Insets, from left: Diatoms feed the copepods that are eaten in turn by the worms that sustain the stickleback fish, pushing toxins up the osprey's food chain.

Diatoms **Copepod** **Worm** **Stickleback**

Genistellospora homothallica

home inside a house fly

inside a

house fly

gut fungi and arthropods share intimate symbiosis

8

Merlin White

Left:
Rhizopus stolonifer, or pin mold, beards a rotting tomato. The mold derives from the ancient Zygomycota phylum in the kingdom of Fungi. Zygomycetes blacken bread and ferment soy sauce.

Below:
Black flies and gut fungi share a lopsided but mostly benign commensal symbiosis that benefits the endosymbiont without harming the host. Previous: fertile thallus tips of *Genistellospora homothallica*, a gut fungus from black flies.

On a warm, spring day in May, Merlin White, a friendly and upbeat associate professor of biological sciences at Boise State University, takes three of his students down to the Boise River to hunt for Trichomycetes, or gut fungi, a unique group of microbes that inhabit the digestive tracts of aquatic insects, among other critters. First, the group heads over to a locker room where they don chest waders and rubber boots. Then, they grab some sampling materials and a small cooler and take the greenbelt pathway over to the Boise River in front of the Morrison Center.

White walks a few feet into the shallow water close to shore and starts kicking over rocks, sticks and woody debris, stirring things up. He dips a rectangular plastic tub into the water to capture some small aquatic insects in the larva or nymph stage. Specifically, they are hoping to collect some midges and mayflies, which may be inhabited by gut fungi. In minutes, White has a number of active live candidates in his container; the translucent creatures twist and turn and squirm. He takes an eyedropper, captures them, drops them into a smaller plastic container and puts a lid on it.

The students, Nicole Reynolds, Mason Hinchcliff and Tyler Pickell, gather samples from other parts of the river nearby in the shallows. It doesn't take long. With plastic trays full of tiny aquatic creatures, the students sit on river cobble next to the river, capture aquatic midge and mayfly larvae with eyedroppers, and drop them into smaller containers that are placed in the cooler and brought to the White Mycology Lab for further identification.

"Oh my, I'm really happy to get all of these samples," a beaming White says. "It's easy pickings today."

The group climbs two stories of stairs to reach White's lab in the science building and go to work. White sits in

U.S. Department of Agriculture

Right:

The fungal parasite *Rhopalomyces elegans* bores through the eggs of nematode worms. Australian artist Jenny Manning puns on the rope in *Rhopalomyces*.

front of an Olympus SZ60 microscope, places a mayfly in water on a slide and begins to dissect it. He uses a jeweler's forceps and works quickly and carefully to remove its exoskeleton. He isolates and extracts a tube, the digestive tract. "With these tubules, you can distinguish the midgut from the hindgut regions as you move from the beginning to the end of the gut," he says. "Inside the midgut of some hosts, there is a lining that's kind of like a conveyor belt for the ingested food particles."

The target species, gut fungi—if they are present—hang out at the tail end of the tube, in the hindgut, sometimes even near the anus. White carefully dissects the hindgut, using the fine-tipped instruments, and is pleased to uncover a reputed new species of *Smittium*, one that they had not seen for a couple years, in midges from the Boise River. "Oh, this is a beautiful specimen," he whispers while looking through the microscope. "I can't believe we found this again."

He puts the gut fungus specimen on a slide and moves to a more powerful microscope, a Nikon Eclipse 80i. "Oh my, this one has beautiful asexual spores," he says. Under stronger magnification, the gut fungus resembles a shrub with branches, or perhaps more precisely, a chandelier with lit candles. On the tips, what look like candles are *sporangia*, oblong containers of spores, which are ready to spread into the stream environment as the mayfly approaches maturity. "These are the asexual spores," White explains. "At maturity, these trichospores will detach and pass through the hindgut and beyond the anus and into the stream environment until they are taken up or ingested by a suitable host."

Jenny Manning

David Hughes/Penn State University

Since White arrived at Boise State University in 2007, he and his students have been collecting hundreds of specimens of gut fungi from host species in the Boise River, Cottonwood and Dry creeks in the Boise foothills, and other streams nearby in the Boise National Forest to document existing and new species. Up to the time that White arrived at Boise State, no one had studied these microbes anywhere in Idaho. The state was ripe for the picking for White, who earned his Ph.D. in botany (mycology) at the University of Kansas and studied under the world's foremost authority on Trichomycetes, Robert Lichtwardt.

"It's ever fascinating," White says. "In the new millennium, we can step into rivers and creeks in the Boise Valley and see new species. We've already discovered five new species in Boise since 2007. I was very fortunate to work with Bob on my Ph.D. He was an incredible mentor," White continues. "Most of us who are involved in the study of Trichomycetes have been inspired by his work."

Indeed, in *Mushroom: The Journal of Wild Mushrooming*, an author interviewing Lichtwardt said this by way of introduction, "It's not everyone who can say that they are the authority for an entire class of fungi, but Robert Lichtwardt can make exactly that claim. By means of a steady stream of publications starting in the early fifties (1950s), he has established himself as the world's preeminent authority on the Trichomycetes."

And now, White, who has sampled Trichomycetes in select locations around the globe, is becoming a world authority himself. The graduate students who work under White have an opportunity through field work, re-

Left:

Ophiocordyceps unilateralis, nicknamed the zombie fungus, spikes from the skull of an ant. The parasite directs the host to a leaf in a humid forest where conditions are ripe for the release of fungal spores. Brainwashed, the ant clamps onto the leaf and dies.

Below:

Ants are arthropods with external skeletons and jointed limbs. From enslaved zombification to subterranean gardens of fungus grown for food, the ant-fungi interdependence can be fickle and complex. Pictured: sharp mandibles of a leafcutter ant.

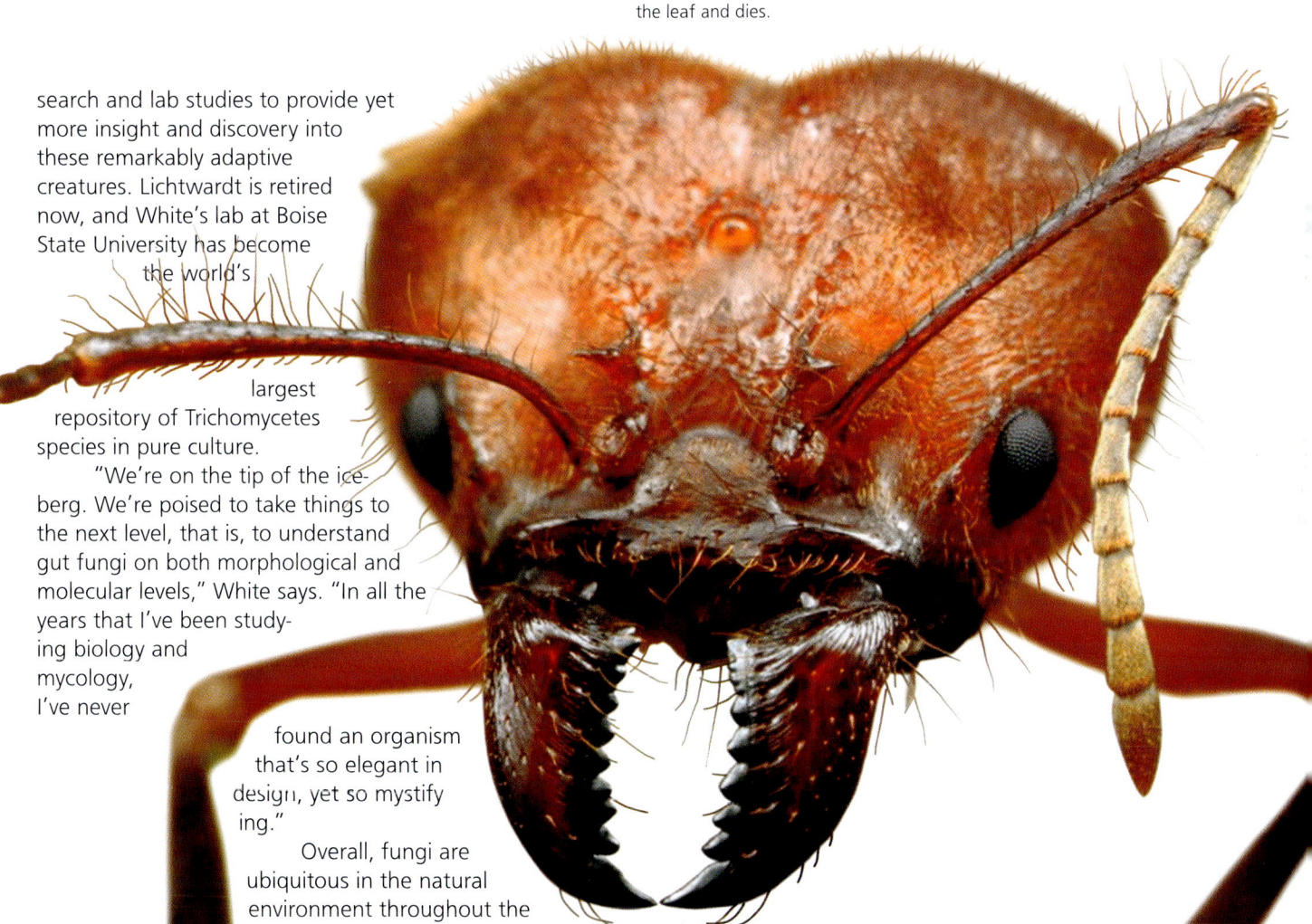

search and lab studies to provide yet more insight and discovery into these remarkably adaptive creatures. Lichtwardt is retired now, and White's lab at Boise State University has become the world's largest repository of Trichomycetes species in pure culture.

"We're on the tip of the iceberg. We're poised to take things to the next level, that is, to understand gut fungi on both morphological and molecular levels," White says. "In all the years that I've been studying biology and mycology, I've never found an organism that's so elegant in design, yet so mystifying."

Overall, fungi are ubiquitous in the natural environment throughout the

Alexander Wild

gut fungi 191

American Philosophical Society

world. They are members of a kingdom of eukaryotic organisms—microorganisms with a nucleus—that include yeast, molds and mushrooms. Whereas mushrooms and molds are more familiar to the average person, other species of fungi such as Trichomycetes are lesser known because they're out-of-sight, out-of-mind tiny organisms that are tucked away inside the digestive tracts of their hosts. They weren't even discovered until the mid-1800s, but based on DNA evidence, they may have been around as long as their host species, the insects, and presumably have co-evolved with them over the past 200 to 250 million years.

A citizen scientist and medical doctor, Joseph Leidy, discovered what he perceived to be a type of fungus in the digestive tract of a millipede collected in his home state of Pennsylvania, reported in 1849. He described it as "intestinal moss."

"He opened the gut and said, 'There's a forest in here!'" White says. "He saw a forest of diversity and density of material. His color drawings of these organisms were beautiful!"

Leidy discovered a species that was later determined to be part of the general category of protists, but nevertheless, this forest of microorganisms that he saw under his microscope, including free-living bacteria, no doubt assisted in digestion for millipedes. Beginning in 1905, two French protozoologists, Louis Léger and Octave Duboscq, delved into the study of marine crustaceans, millipedes and aquatic beetles, and found more that would be considered protists as well. At the time, Léger and Duboscq thought the species were some type of fungus.

"There was limited knowledge at the time," White says. "If you were a fungus, you got dumped onto the plant side of the tree of life."

In 1929, Léger and Duboscq discovered the first species of true gut fungi, *Harpella melusinae*, inside the midguts of black flies. This species is

Left:
Joseph Leidy of Philadelphia (1823–1891) was, said his biographer, "the last man who knew everything" about natural science. A founder of parasite science, or parasitology, he was among the first to describe the fungi and protozoa in the guts of arthropods.

Below:
Millipedes featured large in Leidy's research. His 1849 address to the National Academy of Sciences spoke of the insect's fuzzy microbial forest of "intestinal moss" Inset: engravings from Leidy's monumental *Fresh-water Rhizopods of North America* (1879).

classified in an order of Trichomycetes called Harpellales, which live in the guts of aquatic insects in the larva or nymph stage. Three years later, Raymond Poisson, discovered the genus *Asellaria* in the second order of fungal Trichomycetes. The Asellariales reside in the digestive tracts of adult terrestrial, aquatic, and marine isopods (crustaceans) or springtails.

Five more decades of research brought on the discovery of more species of protists and gut fungi. In 1986, Lichtwardt published the first monograph on Trichomycetes at the University of Kansas and established a formal classification for all the known species. But questions remained about their precise relationships. The word *Trichomycetes* comes from a Greek word, meaning "hairlike fungus," owing to their appearance along the guts of some dissected specimens.

Scientists have discovered gut fungi in the digestive tracts of black flies, mosquitoes, stoneflies, midges, mayflies, crustaceans, springtails and more

gut fungi 193

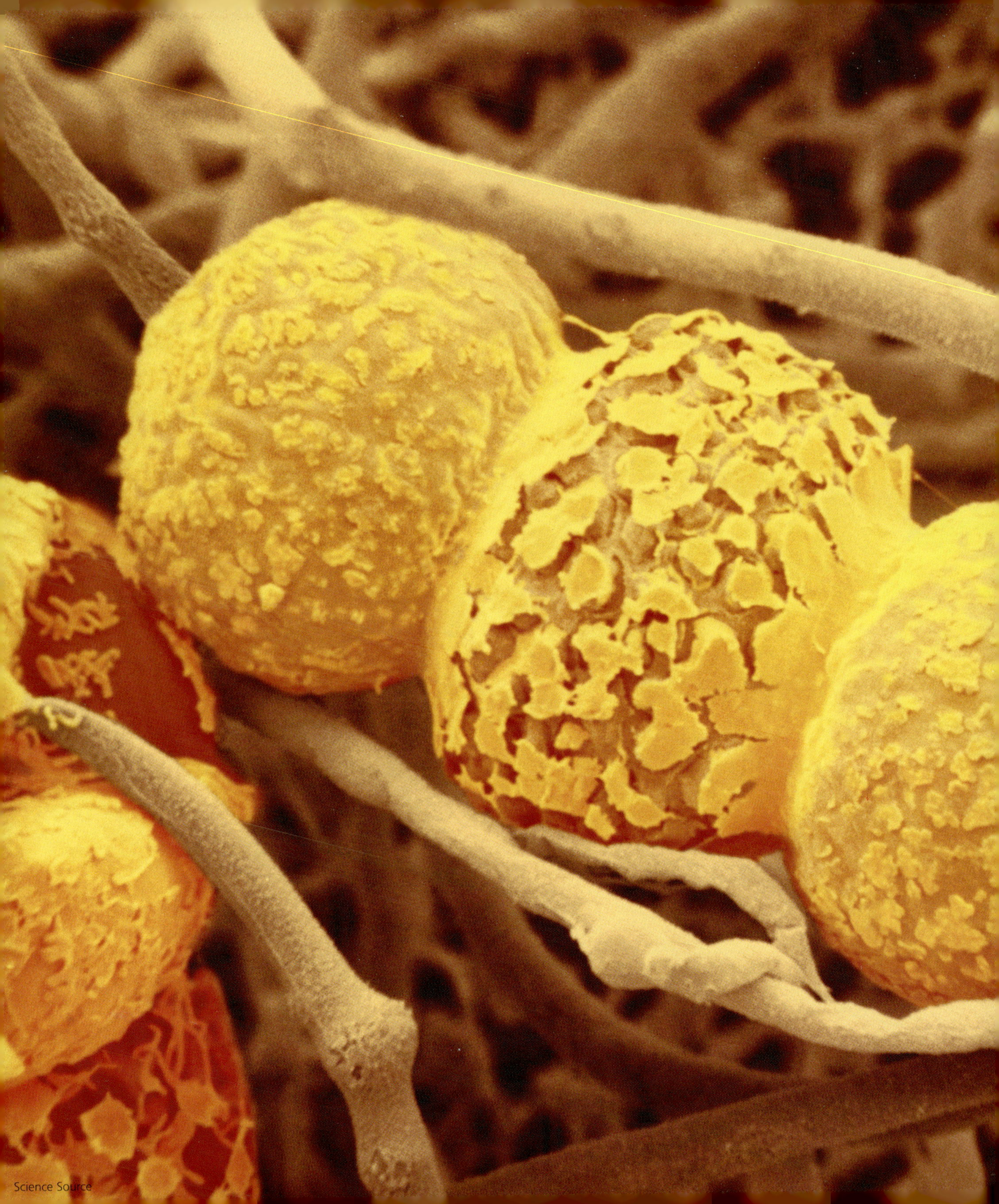

Science Source

Left:
Trichomycetes reproduce both sexually and asexually. In the sexual phase, the hyphae fuse male and female cells, giving birth to a zygospore. Sexuality enhances genetic variations, helping parasites keep pace with their host.

Below:
Intimacy can become win-win mutualistic when parasites provide a vital service. The bobtail squid, pictured, boards a luminous bacterium that glows like a flashlight. The microbe allows the squid to hunt in the darkness of night.

worldwide. There are more than 200 species of Trichomycetes, more than 350 species if you include the protists. All of the hosts of Harpellales and Asellariales species are non-predaceous, or herbivores, eating algae and other plant material, perhaps off the bottom of streams or other moist terrestrial habitats.

"Predatory hosts, including some caddisflies, tend not to have gut fungi," explains Emma Wilson, who recently completed her master's degree at Boise State University under White's direction. "If you look at the gut of a rat or a cat, the guts of predators tend to be simple. But you need a more complex gut structure to digest plant food."

Tricho species have a symbiotic relationship with their hosts and undoubtedly assist in digestion. In biology, a symbiotic relationship means that two or more organisms live in close association with each other. The word *symbiosis* comes from a Greek term, meaning "living together." One species depends on the other for survival. The relationship between Tricho species and their hosts can be mutual (i.e., commensal, meaning the species are helpful to one another or one is neutral to the other) or parasitic, such as with some species of gut fungi that inhabit mosquitoes and black flies.

The species *Smittium morbosum* can be lethal to mosquito larvae: the fungus penetrates the gut, attaches to the inner exoskeleton and prevents molting

University of Wisconsin, Madison

The Gut's Microbiota

The human metabolism is a microbiota abloom with bacteria, protozoa and fungi. More numerous than we can we count, more vital than we can imagine, the microorganisms in our large and small intestines outnumber our human cells by a factor of 10 to 1.

Scientists estimate that 100 trillion organisms of more than 1,000 species inhabit the human digestive system. Bacteria dominate. Storing fat, fighting disease, they first pass from the mother to the sterile gut of an infant via the placenta and breast milk. Some microbes send signals from the gut to the brain, regulating emotions. Researchers at Arizona State University have recently suggested a link between gut microbes and autism spectrum disorder. Because many autistic children suffer from gastrointestinal problems, scientists have speculated that an imbalance of microbes disrupts the central nervous system.

Microbes are also the gut's gatekeepers, blocking pathogens. A dangerous exception is the hardy cyst-forming protozoan that causes amoebic dysentery. Other gatecrashers are autoimmune diseases like rheumatoid arthritis and multiple sclerosis. But some guts are stronger than others. Your body's microbiome, like your fingerprint, is unique.

Pictured: The gut's microbiota includes, from left, *Bifidobacterium* from breast milk; *Lactobacillus*, an aid to digestion; *Salmonella*, an agent of food poisoning; and *Entamoeba*, a source of dysentery. Right: the gastrointestinal tract from mouth to anus is about 10 times longer than the length of the body.

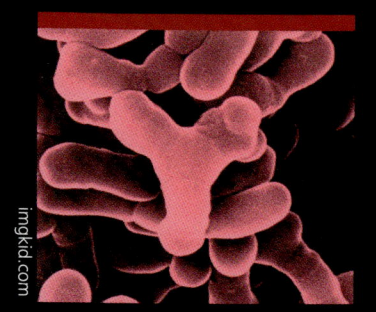

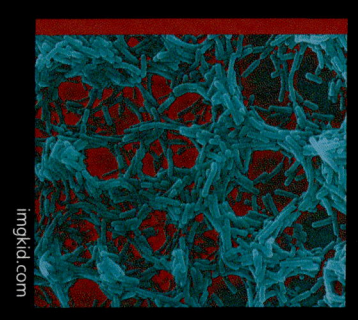

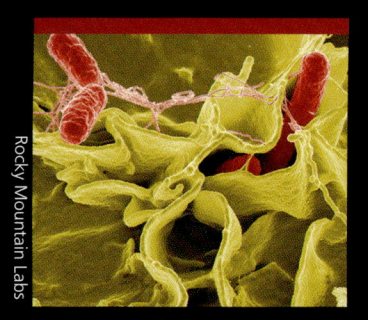

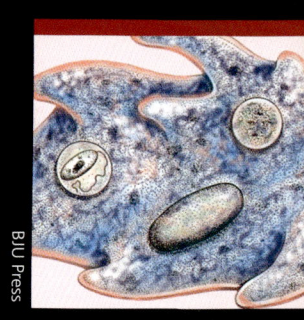

Bifidobacterium | ***Lactobacillus*** | ***Salmonella*** | ***Entamoeba***

Science Source

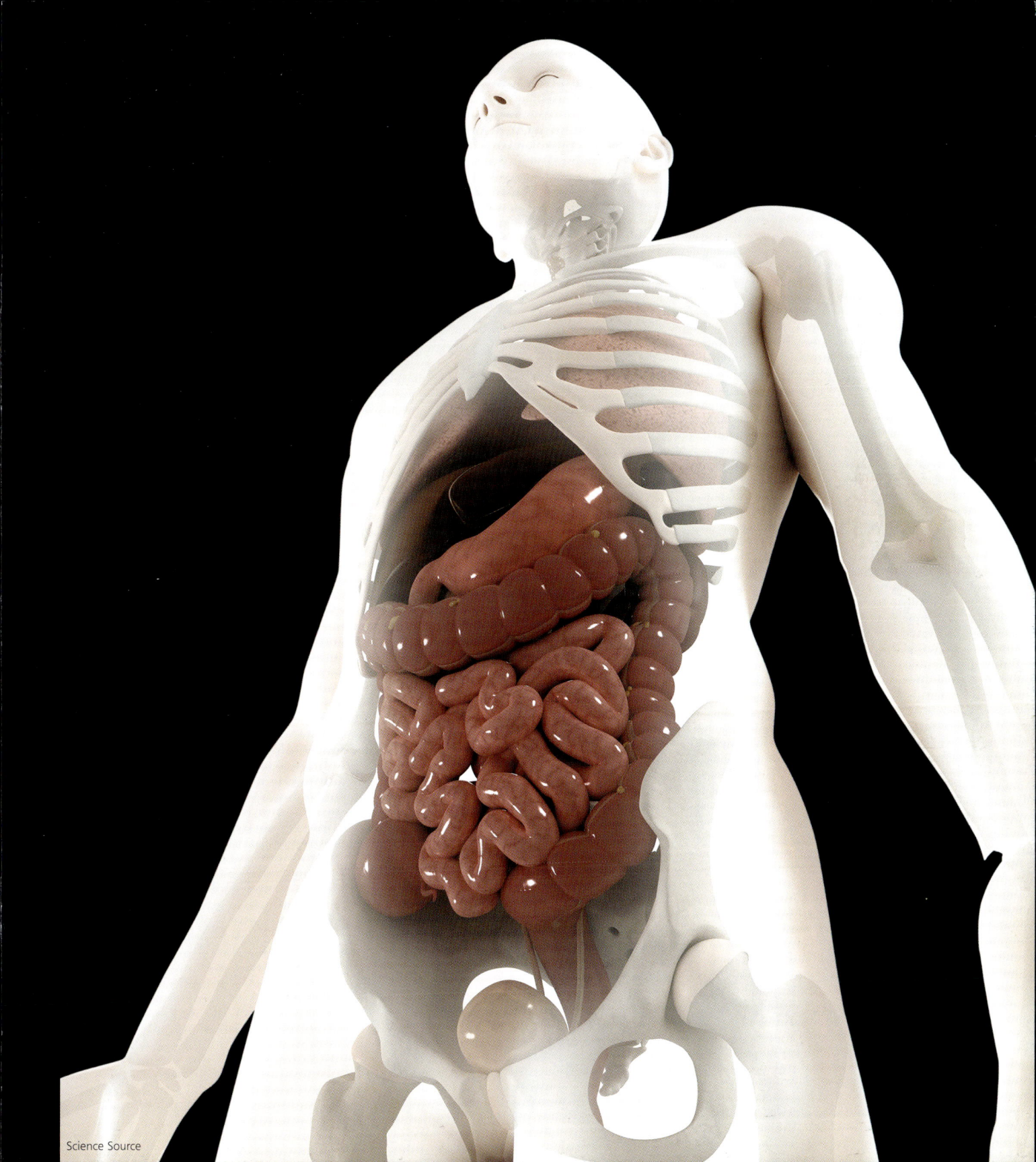

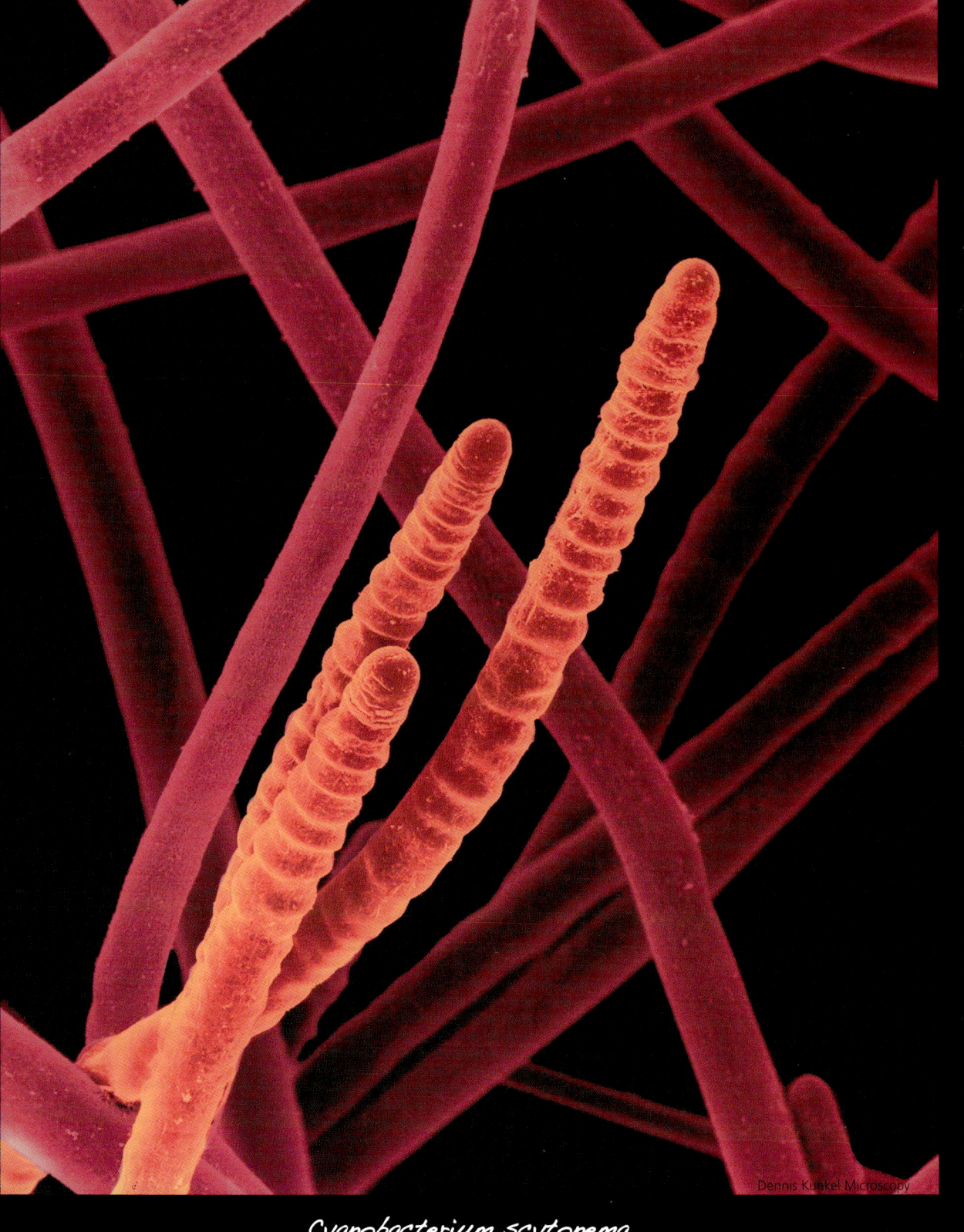

Cyanobacterium scytonema

freddy fungus meets alice algae

biological soil crusts on Idaho's rangelands

9

Chi E Shenam

Left:
Reddish brown *Psora decipiens*, a lichen, patches in the open spaces between clusters of shrubs. Its common name is blushing scale.

Below:
The greater sage-grouse needs low shrubs for shelter. Where an estimated 16 million of the chicken-like birds once populated western rangelands, fewer than 500,000 remain.

them has been emerging for about a decade—owing in part to Rosentreter's scientific papers and books co-written with other soil and plant experts. Soil crusts are becoming part of the conversation now, as scientists figure out what role they play in the composition of healthy rangelands. "There's a small group of us who have known about soil crusts in the scientific community, but the word is just getting out to a larger audience," notes Beth Newingham, an assistant professor of natural resources at the University of Idaho who specializes in rangeland ecology and restoration.

Rosentreter worries that the shallow, rocky soils in low-elevation rangelands bordering the Snake River Plain are losing their integrity in places where the soil crust foundation has been lost as a result of repeated wildfires, the invasion of exotic plants and weeds such as cheatgrass and medusahead wildrye, heavy livestock grazing, off-road vehicle recreation activity, BLM chaining and tillage projects, and other disturbances.

"With soil crusts, the biggest thing is soil binding," he says. "Crusts hold the moisture in the soil, and they protect the soil as a skin on the Earth. Our soils came from volcanic activity in the Cascades—lots of ash and dust. It took 30,000 years to metamorphize that material into productive soil that can support plant and animal life, and it's taken just 50 years to erode it away. That's scary."

John Muir Laws

fungus meets algae 213

Below:
Cyanobacteria concentrate in the top two inches of crust. Opposite: Heat and radiation in deserts with sparse vegetation can darken and redden the soil.

soilcrust.org

Let's keep the soil in place. Let's not turn our desert lands into a Dust Bowl."

Rosentreter retired from the BLM in 2013 after more than 35 years with the agency, so he's free to speak his mind about soil crusts. But he's been working on educating BLM range managers and the general public about soil crusts for some time. With five co-authors, Rosentreter wrote a paper titled *Biological Soil Crusts: Ecology and Management*, published in 2001. This was a technical guide for BLM rangeland managers and ecologists throughout the western United States. "That's where the information congealed about soil crusts," he says.

Rosentreter also co-authored *Biotic Soil Crust Lichens of the Columbia Basin* in 2007, *A Field Guide to Biological Soil Crusts of Western U.S. Drylands* in 2007, and numerous technical papers for the U.S. Department of Interior and BLM. In the lichen guide, Rosentreter and co-author Bruce McCune write, "Botanists and consultants with resource management agencies need to know about soil crusts to better manage grazing, fire and other disturbances. But at present, few botanists can identify the species or even the genera of lichens involved in soil crusts. Even people with training in lichens are often at a loss with soil crusts."

In the introduction to *Biological Soil Crusts: Ecology and Management*, the authors explain, "Biological soil crusts are a complex mosaic of cyanobacteria, green algae, lichens,

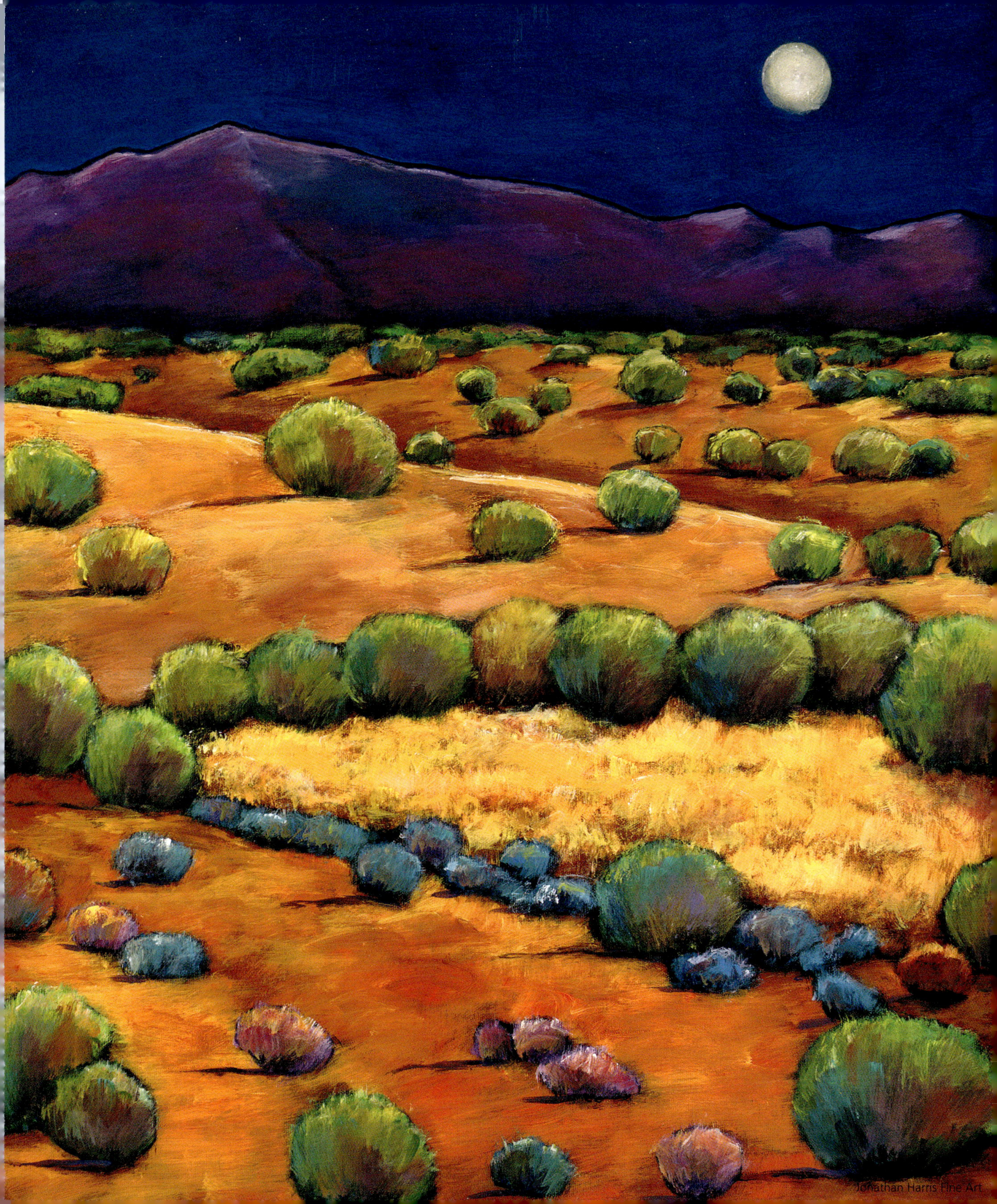
Jonathan Harris Fine Art

Right:
Cheatgrass invades the rangelands where grazing fractures the crust. The invader robs soil of moisture. Range fires burn twice as hot.

writes in *A Field Guide to Biological Soil Crusts of Western U.S. Drylands* with co-authors Matthew Bowker and Jayne Belnap. "They were present in the oceans over 3 million years ago and have existed on land for over 1 billion years."

Cyanobacteria often provide most of the "glue" that holds biological crusts together. Some species bundle filaments with a sticky gelatinous sheath. "These bundles of filaments wind throughout the uppermost soil layers, forming a net-like structure that binds together soil particles. This forms soil aggregates that create pathways for water infiltration and surfaces for nutrient transformations, while also increasing the soil's resistance to wind and water erosion," the authors write.

After the cyanobacteria colonize in the soil, the mosses move into the budding soil crust community, and then the lichens follow. But how? Mosses and liverworts are known as *bryophytes*, very small non-vascular plants. They reproduce by sending out spore capsules that rise above the moss. Inside the spore capsules, known as *sporangia*, haploid spores are produced

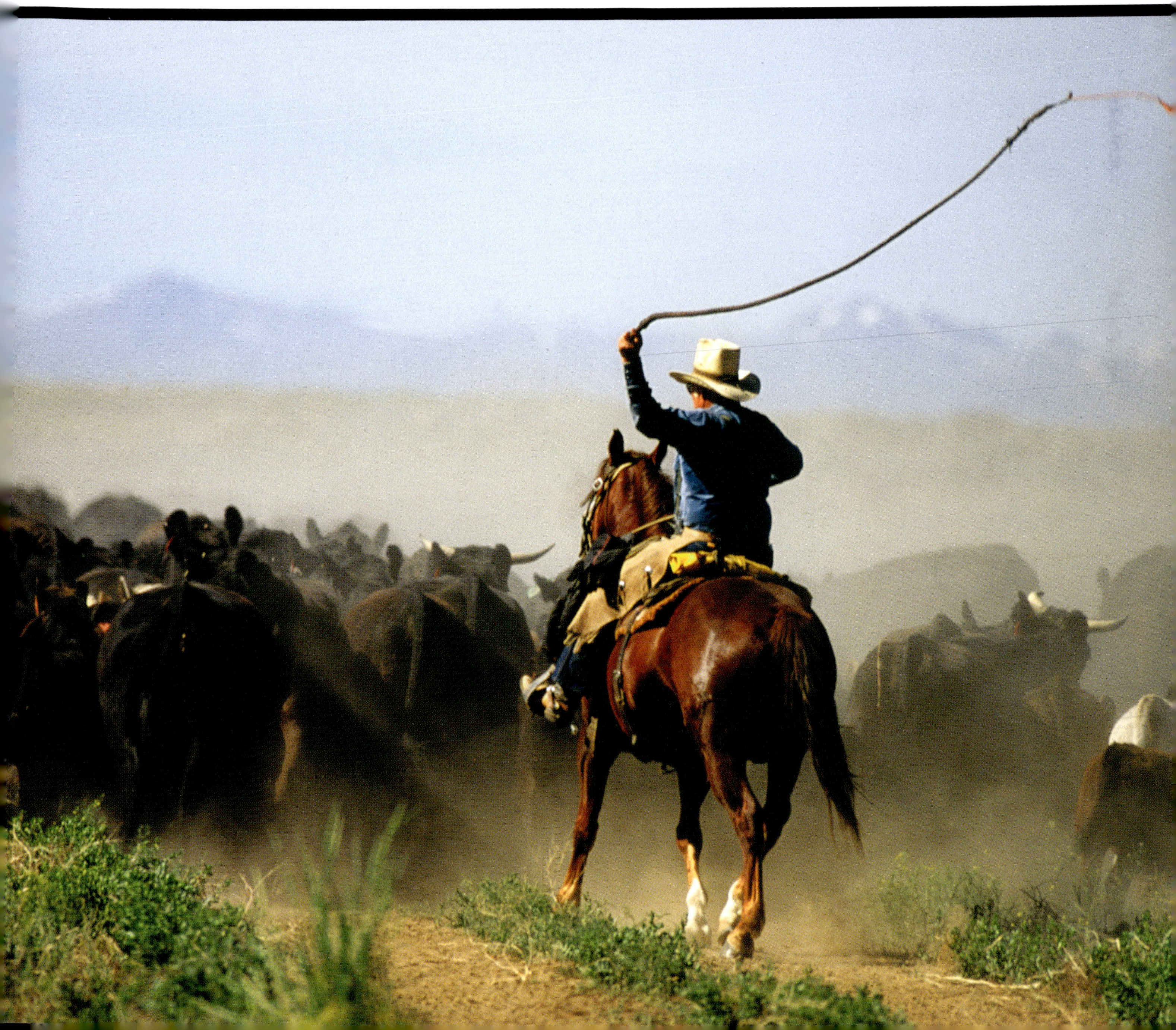

management plans come up for renewal, but data on this topic are lacking.

Perhaps the largest stroke to protect soil crusts in southern Idaho occurred when Owyhee County ranchers worked together with a number of environmental groups and U.S. Senator Mike Crapo's staff to preserve 517,025 acres of wilderness in the Owyhee Canyonlands in 2010. The wilderness areas and wild and scenic rivers are protected for multiple reasons, including preserving scenic values, wildlife habitat, recreation values, natural ecological values and more. The area includes the Little Jacks Creek and Big Jacks Creek wilderness areas, North Fork of the Owyhee area, Jarbidge-Bruneau area, East Fork of the Owyhee area, and Pole Creek, next to Chris Black's private ranch in the Owyhees. The wilderness designation should preserve native landscapes and soil crusts in perpetuity.

"The creation of the Owyhee wilderness areas will really help," Rosentreter says. "I never thought I'd see that happen in my lifetime."

●

Right: Chris Black and his son, Justin, promote holistic sustainable grazing in Idaho's Owyhee County. Above: pronghorn antelope, a shrub-steppe native rescued from the brink of extinction.

George Booruhy

Molly Messick / StateImpact Idaho

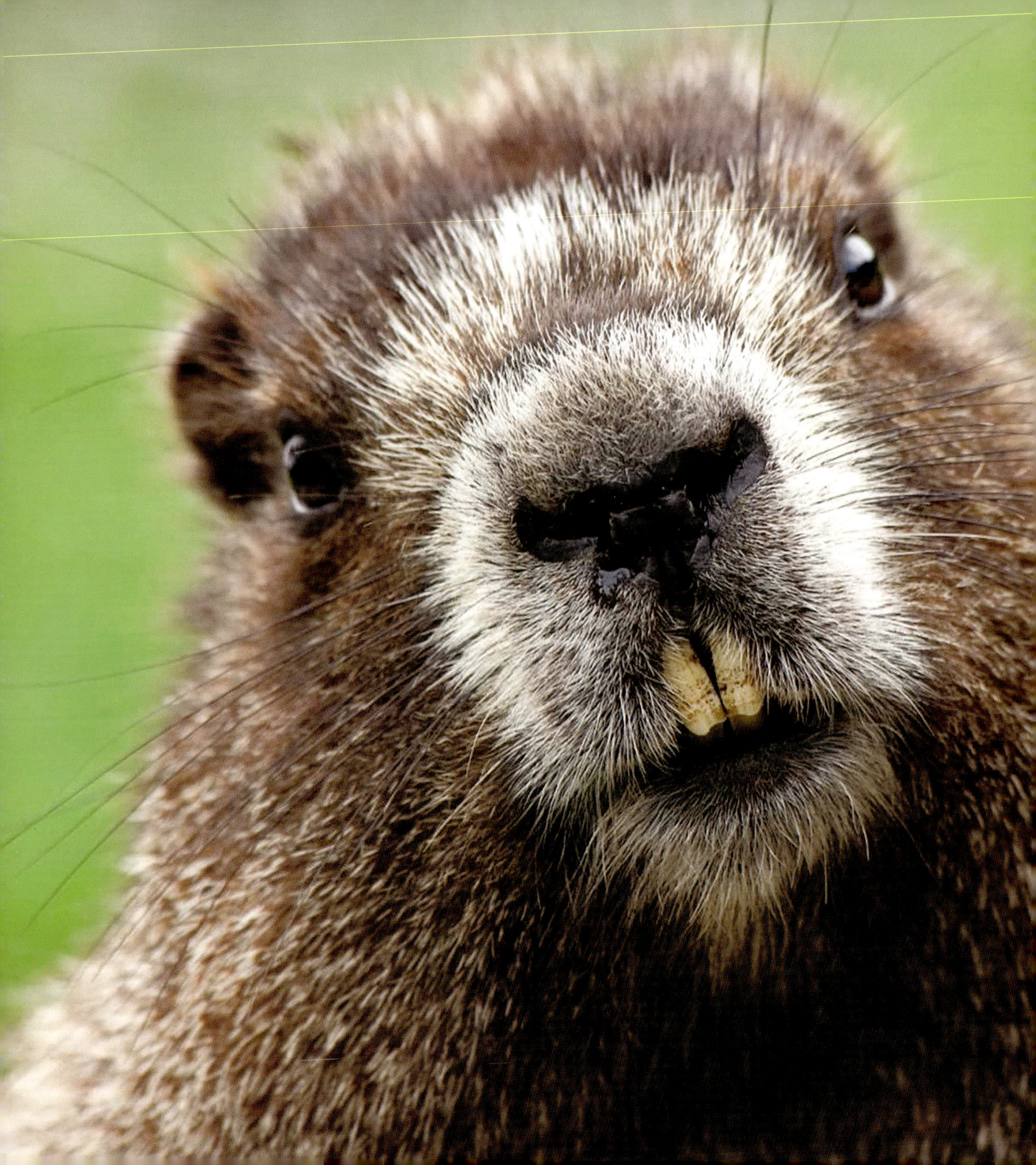

Leave It to Beavers

Beavers are the wetland engineers that build habitat for many species. Beavers, with their tunnels and dams, trap pollutants and raise the streams for deepwater organisms. When dams break, flooding sediment, the beaver ponds recharge the wetlands with microscopic zooplankton. Microbes reward the rodent by giving beavers their talent for digesting bark and wood. A pouch at the top of a beaver's upper intestine houses bacteria and protozoa that break down cellulous fiber in wood. It can take 40 hours and several passes through the beaver's digestive system to metabolize fibrous food.

Pictured: The beaver (*Castor canadensis*) spreads microbes at the base of the food chain. Insets: freshwater zooplankton; raising the stream, pooling the wetlands.

Zooplankton   **Beaver Engineering**

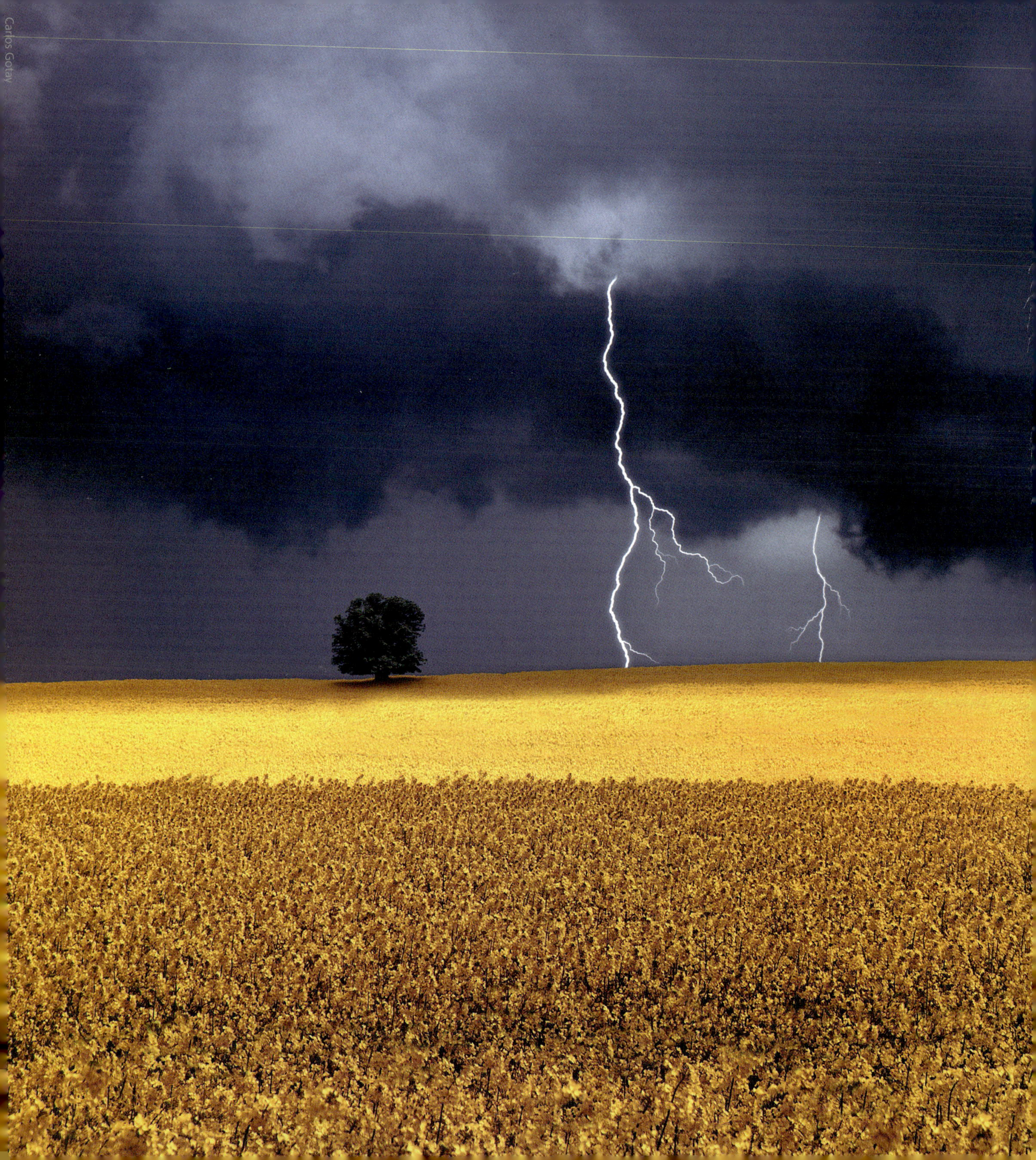

Carlos Gotay

saving the soil

Idaho farmers embrace the no-till revolution

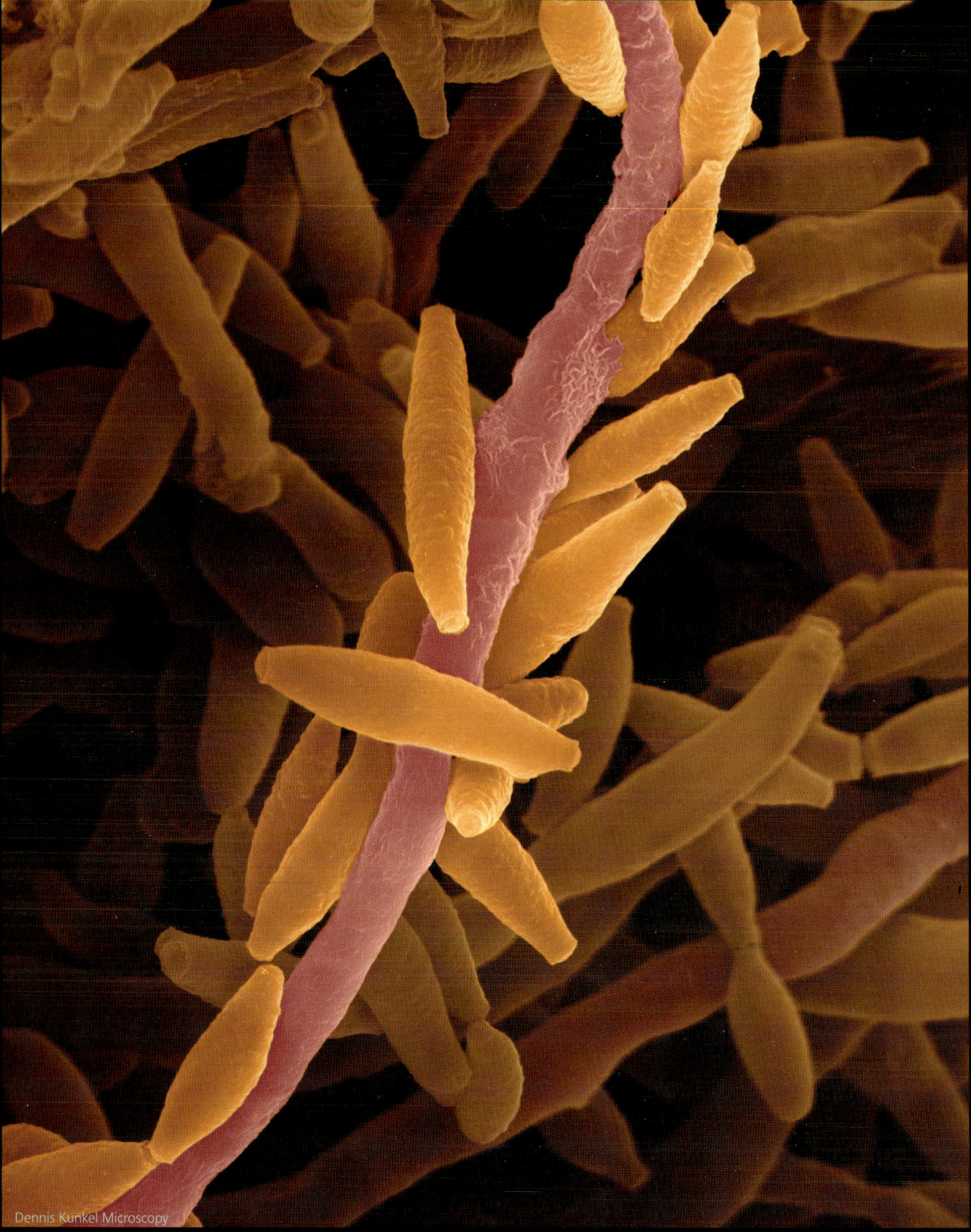
Dennis Kunkel Microscopy

Left:
Fungal hyphae and spores weave a sticky fibrous web for the topsoil's microflora. Previous: Canola, or rapeseed, provides summer cover for winter wheat on the golden Palouse of Idaho's Camas Prairie.

Below:
Black storms of dust swept cropland in the 1930s. Congress responded with the Soil Conservation Service, created in 1935. Pictured: harvesting beets on land reclaimed from the Dust Bowl.

Marlon Winger grew up on a farm near Preston in southeast Idaho. His family raised alfalfa, beans, barley, peas, sugar beets and corn. They farmed the traditional way—tilling the soil in the spring to bury the weeds, fluff up the soil and plant the year's crops. Then they applied fertilizers to grow bigger yields and sprayed the fields with farm chemicals to keep weeds, pests and fungi at bay.

Winger liked farming, so he studied agriculture as an undergraduate and got a master's degree in soil science from Utah State University. He's been a career agronomist for the Natural Resources Conservation Service (NRCS) for about 25 years. He helps farmers raise the best crops possible while taking care of the land. A few years ago, Winger attended an NRCS soil health workshop in Bismarck, North Dakota, and had an epiphany after visiting Gabe Brown's farm.

"It made me a believer," Winger says. "We saw his soil. It was beautiful, rich with nutrients and earthworms. I smelled it. I touched it. This is not a fad. The science is there behind everything they do. It's not some magic foo-foo. They're trying to farm in nature's image."

The Browns switched to no-till farming 20 years ago, moving to a system of "holistic management." Instead of tilling the soil multiple times during the year, Brown leaves the soil alone to allow a natural community of microorganisms to populate under his crop fields and livestock pastures. He uses cover crops to add diversity to the microbial population in the soil and stop soil and wind erosion. He has reduced the use of commercial fertilizer by 90 percent and herbicides by 75 percent. "These strategies have allowed the health of the soil, the minerals and water cycles to greatly improve," Brown says. "This results in increased produc-

Mike Edminster

tion, profit and a higher quality of life for us. We are moving toward sustainability for not only ours but future generations as well."

Winger talks non-stop as we drive to Kuna to look at some farm fields. He's a busy guy who's constantly on the run, giving soil health workshops to farmers throughout Idaho. Driving along on a clear, summer morning, Winger passes by crop fields, irrigation canals and ditches in an agency SUV. He sud-

Left:
More than a third of the nation's croplands are direct-seeded with minimal tillage, according to the USDA. No-till converts boast cost savings in fertilizer, labor and fuel.

Below:
Root vegetables thrive in no-till gardens where cover crops replenish the soil. Inset: Marlon Winger, state agronomist for the USDA's Natural Resources Conservation Service.

denly brakes and we pull over by a large pile of dirt next to a canal. This is a sediment trap at the end of an irrigation ditch. The big pile of dirt is the topsoil lost from the farm fields next to the ditch from furrow irrigation. The sediment trap keeps most of the dirt from flowing into the canal, and eventually the Boise River, but to Winger, "it's like putting a Band-Aid on a heart attack. It's not very sustainable to keep washing off the topsoil," he says.

We stop by Winger's house in Kuna, where he shows me a no-till garden, a few chickens and sheep. He uses mulch and manure to fertilize the garden, and you can't see any bare soil anywhere. He keeps it all covered to prevent weeds from growing and feed nutrients into the soil. He takes a shovel, digs into the soil and turns it over. The deep-brown soil looks porous, and even though it's been more than a week since the last rain, earthworms are wiggling around in the dirt. "I haven't tilled the garden for several years," he says. "The soil is getting healthier all of the time."

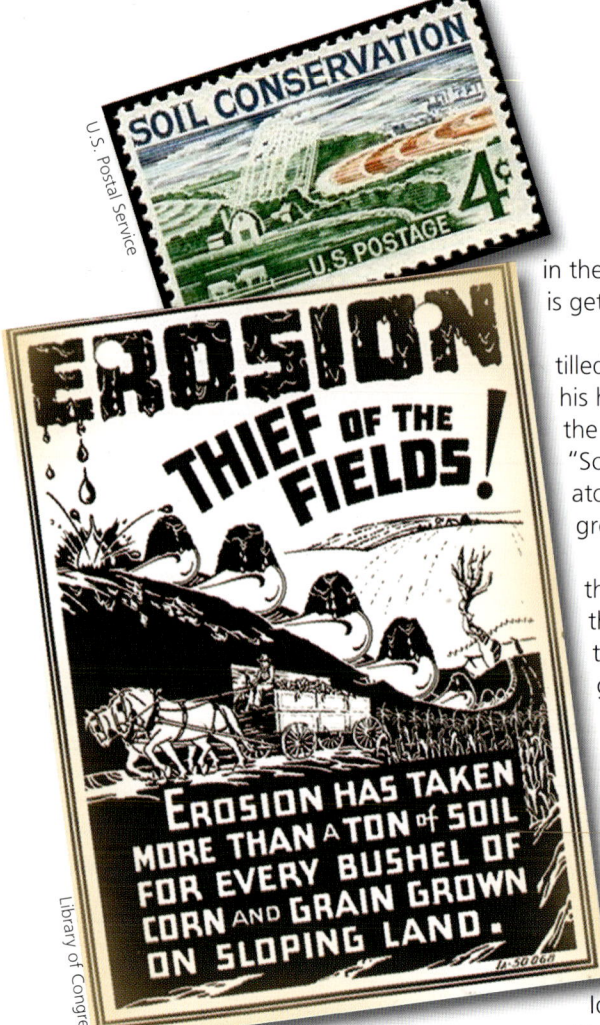

Then Winger demonstrates what happens to the soil when it's tilled. Standing over a field of freshly tilled dirt, he raises his shovel over his head and slams it into the ground. "We can't continue to pulverize the soil," he says. Suddenly this mild-mannered fellow is on a tear. "Some people say that tilling the soil is the equivalent of dropping an atomic bomb on the farm. It destroys the microbial community that's growing in the soil."

But to convince farmers to change the way they've been farming their whole life is hard. "We all grew up thinking that the more we till, the more we fluff up the soil, the better—like roto-tilling the garden in the spring," Winger says. "It's almost like trying to change your religion!"

Steve Riggers has been no-till farming in the Camas Prairie near Grangeville, Idaho, for more than 20 years. He agrees that the cultural paradigm is to till the soil. "There's 10,000 years of culture and ways of doing things behind tillage," he says. "Asking farmers to change their way of doing things is a pretty big leap of faith. But it's catching on ... it's a wonderful way to farm."

About 80 percent of the farmers in the Camas Prairie area are now practicing no-till farming. "It's not an easy deal. There's a lot to learn," says Kevin Seitz, NRCS district manager in Nezperce, Idaho.

Eileen Rowan, a water quality conservationist for the Idaho Soil and Water Conservation Commission in Nezperce, recalls that in the 1980s, the word on the street about no-till was "no till, no crop."

Left:
Erosion was "a menace to the national welfare," said Congress in the wake of the 1930s Dust Bowl. The U.S. Farm Securities Administration responded with soil conservation campaigns.

Below:
A gram of healthy soil can harbor a diverse microflora with a billion organisms. Cyanobacteria, pictured, metabolize the soil's organic matter into nutrients that plants absorb.

The turning point for Riggers came in 1994, when he was about ready to quit farming. "It was sheer drudgery," he says. Using the conventional tillage system, Riggers had to drive the tractor over a 700-acre farm field seven times, applying seed, and then fertilizer, farm chemicals and more. He wanted to figure out a way to be more efficient. He knew a few farmers in the Camas Prairie who had gone to a no-till, direct-seeding system, and he decided to try it.

"It brought the joy of farming back for me," he says. "You're not doing this senseless plowing over and over. Tillage is not good for the resource. It's been a whole new frontier for me."

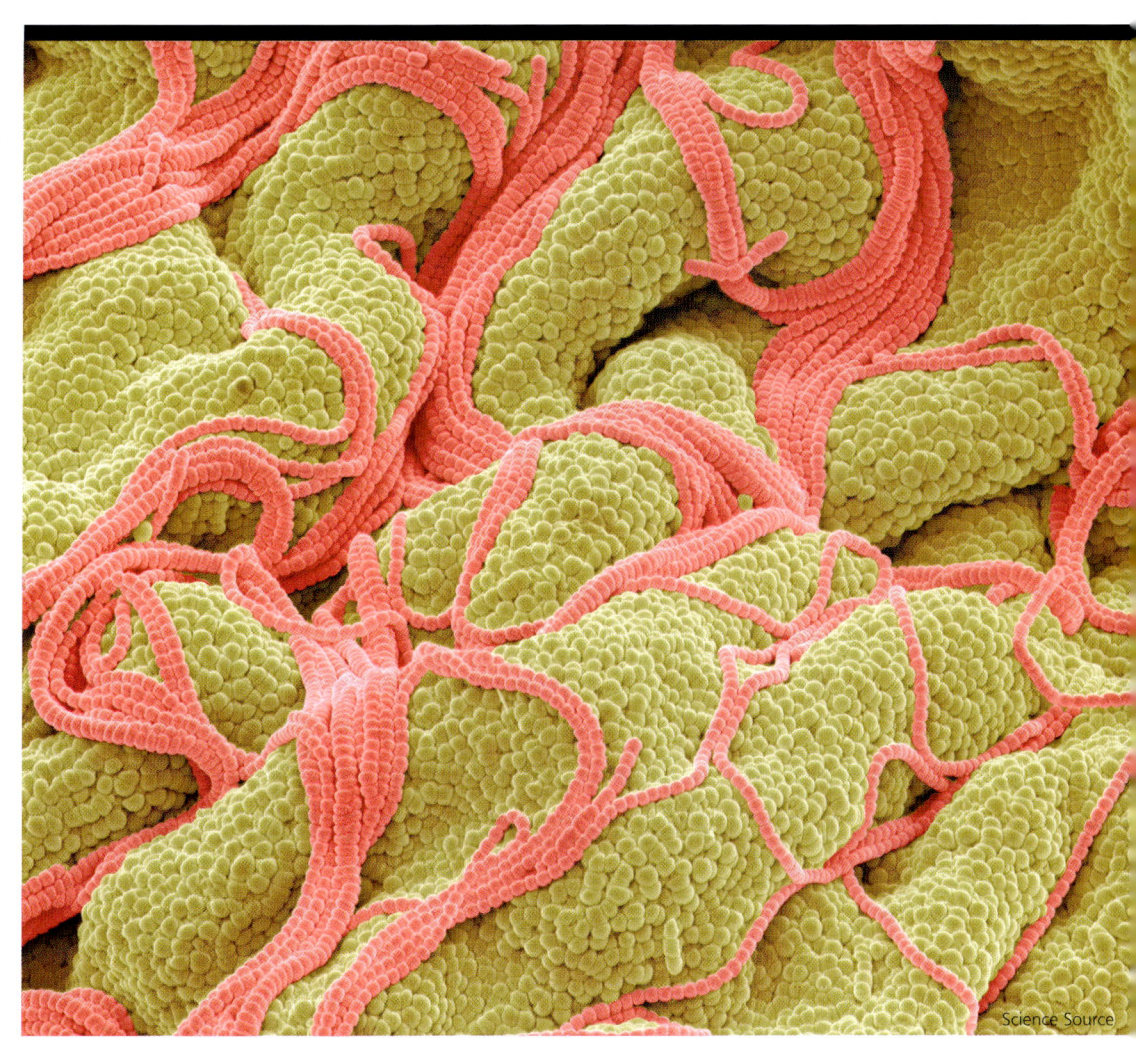

David R. Frazier

Left:
Corn sowing machines use runners, or drills, to plant seeds at a uniform depth. Drills preserve topsoil, sparing damage from tractors and plows.

Below:
Horse-drawn seeding machines date back to 18th-century England. Americans added hollow spikes to deposit the seed. Pictured: American Seeding Machine, 1903.

Riggers farms 6,700 acres in the Camas Prairie area with his brother Nathan. They own some of the farm ground and they rent the rest, a common practice in these parts. They raise several varieties of winter wheat—their best cash crop—as well as spring wheat, green peas, lentils, malting barley and canola. To make the switch to no-till farming, Riggers had to invest in a larger, 500-horsepower tractor to pull a new direct-seeding harrow. The investment cost about $700,000 for the new tractor and drill.

Under the no-till system, he makes just three passes over his fields per year. "We found out that it's easy to direct seed," he says. "We cut our fuel bill by 40 percent." To do the no-till, direct-seed farming, you have to farm "smarter," Riggers says. "It takes more management and planning, but it costs less."

Riggers takes me to a farm field that hasn't been tilled since 1987. It's a mid-July morning in the Camas Prairie, and the sun shines on a field of green canola. The field abuts blond wheat crops, dark-green alfalfa and light-green peas. It's a patchwork quilt of varying colors, all neatly arranged in big squares and rectangles. Even in fields that have been harvested, the soil is covered by residue or a cover crop. You don't see bare soil anywhere.

Canola seeds are used to make canola oil, which used to be a break-even crop, financially, Riggers says, but a good rotation crop for the soil. Because of the increasing

Richard Meredith Browne/Avenue Art

demand for healthier cooking oils like canola, it's more of a cash crop now. "It's a heart-healthy oil," he says. "There's a high demand worldwide for these seeds."

Through trial and error, Riggers and his brother have figured out a five-year crop rotation system that works best for the soil and for their farming business. They grow winter wheat two years in a row, and then they switch to spring cereal, canola and lentils for the other three years. "You have to rotate your crops up here to take care of the pathogens in the soil," he says. "There's a ton of pathogens that you're watching out for. By rotating the crops, you head them off before they become an issue."

Underneath the four-foot-high canola crop, at ground level, the trimmed wheat stubble from last year's wheat crop is still standing a few inches high. The direct-seeding equipment allows Riggers to drill the seed of the canola into the field without tilling the soil. But he still has to apply Roundup to fight off weeds after seeding the canola. To keep his herbicide treatments to a minimum, Riggers plants Roundup-ready canola, made by Monsanto Corporation. It's the only genetically modified crop that they raise, he says, "but we're able to manage the weeds better, and we get higher yields."

When he's tending to his fields, Riggers likes to dig into the soil and look for earthworms. "They're our indicator species," he says. "They have to have dead plant material, crop residue, to survive. That's what they eat. With tillage, you destroy their habitat."

Under the no-till system, it takes about three to four years for earthworms to show up, he says. "They create a moist, rich environment. They allow the water to slowly percolate into the soil, and they help retain that moisture."

Left:
Night crawlers and other deep-burrowing species of earthworms dig tunnels that irrigate plants at their roots. A cubic yard of organic soil can host 300 worms.

Below:
Gordon Gallup of the Idaho Wheat Commission has direct-seeded his farm near Ririe since 1985. Inset: color-enhanced scan of a soil nematode in leaf litter.

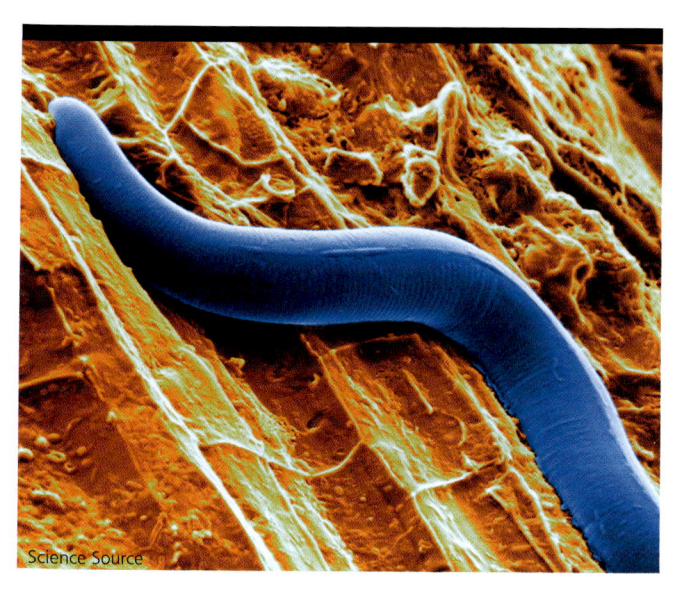

Riggers remembers a three-day, tri-state conference in 1996, during which Roger Veseth, a University of Idaho professor and conservation tillage specialist, shared the results of some experimental crop plots in the Camas Prairie. Veseth did a spring seeding with four different treatments: one field was direct-seeded on top of the wheat stubble; the second field was plowed and seeded conventionally; the third was seeded on top of burned residue; and the fourth was seeded with minimum tillage.

"There were four long strips with each treatment growing alongside each other," Riggers says. "Roger predicted that the no-till would do the best. And it turned out, it was true. It had the biggest yield."

A lot of local farmers attended the conference. "It really kicked off this push into direct-seed farming," he says. "It was one of

puts a chunk of soil in a wire basket at the top of each tube. He pours water over the soil, and we watch the results. The tilled soil falls apart in seconds. We watch the tiny particles of soil dissolve and fall out of the cage into the tube. The no-till soil holds together for minutes on end, absorbing the water, the glomalin at work. "You see? This is how Mother Nature works," Winger says.

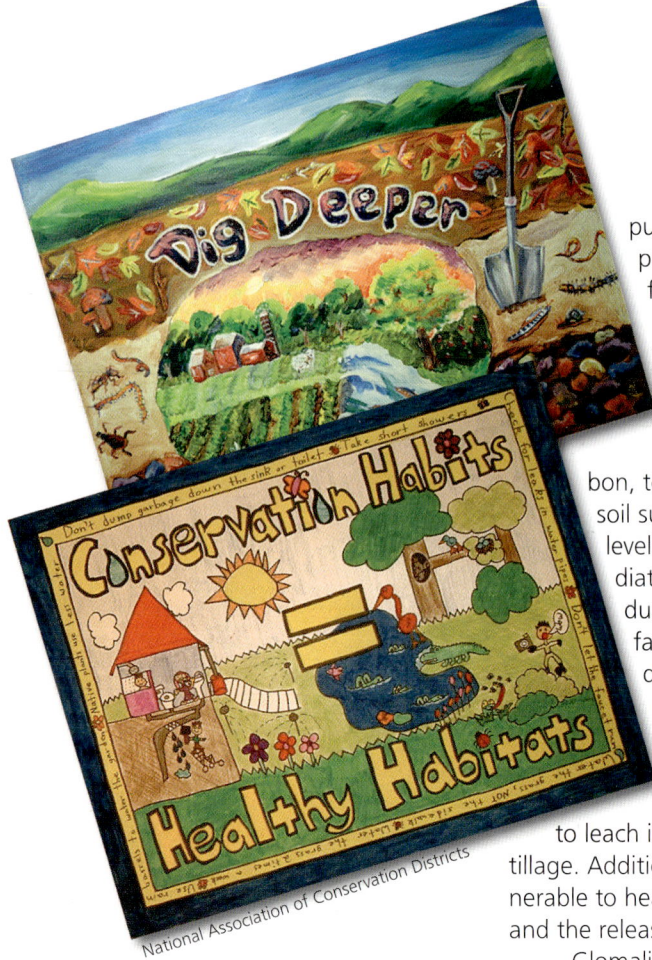

Soil studies show that no-till farming increases organic carbon, total nitrogen and microbial activity in the surface layer of the soil substantially. Tillage, on the other hand, increases carbon dioxide levels, net mineralization and nitrate accumulation in the soil immediately after the fields have been plowed. Tillage disrupts and reduces microbial populations in the surface levels of the soil, but farther below the soil profile, where the tillage equipment has not disrupted the microbial community, the populations of bacteria, fungi and protozoa are still high.

Scientists say the primary and secondary impacts of tillage—particularly the loss of soil strength and moisture retention at the surface level, as well as a higher potential for nitrate to leach into the groundwater—should be balanced with the benefits of tillage. Additional secondary impacts of tillage include leaving farm fields vulnerable to heavy wind erosion, topsoil eroding as a result of surface irrigation, and the release of carbon into the atmosphere.

Glomalin stores a lot of the carbon in the soil. When the soil is tilled, it releases significant amounts of carbon into the environment, adding greenhouse gases to the ozone. "Glomalin may account for as much as one-third of the world's soil carbon—and the soil contains more carbon than all plants and the atmosphere combined," write Mark Amaranthus and Bruce Allyn in the

Left:
"Dig deeper" became the catchphrase for the a poster contest sponsored by the National Association of Conservation Districts. The K-12 program stressed backyard stewardship of sustainable soil.

Below:
Endospores, stained pink, weave like a beaded string through soil-dwelling *Sporosarcina ureae*. Endospores defend against starvation and stress in pastures crowded with cows.

June 2013 issue of *The Atlantic*.

That said, is no-till, direct-seed farming a panacea for everyone? Although it's becoming very popular in some parts of Idaho, such as in the Camas Prairie, it's still a minority practice in southern Idaho, where tillage is still king. Winger estimates that only about 10 percent of Idaho's farms are practicing no-till techniques. "The problem is we're so entrenched in our paradigm that it's tough to change," he says.

Another problem is no one has figured out how to use the no-till technique for growing potatoes—one of Idaho's largest cash crops. Spuds are grown

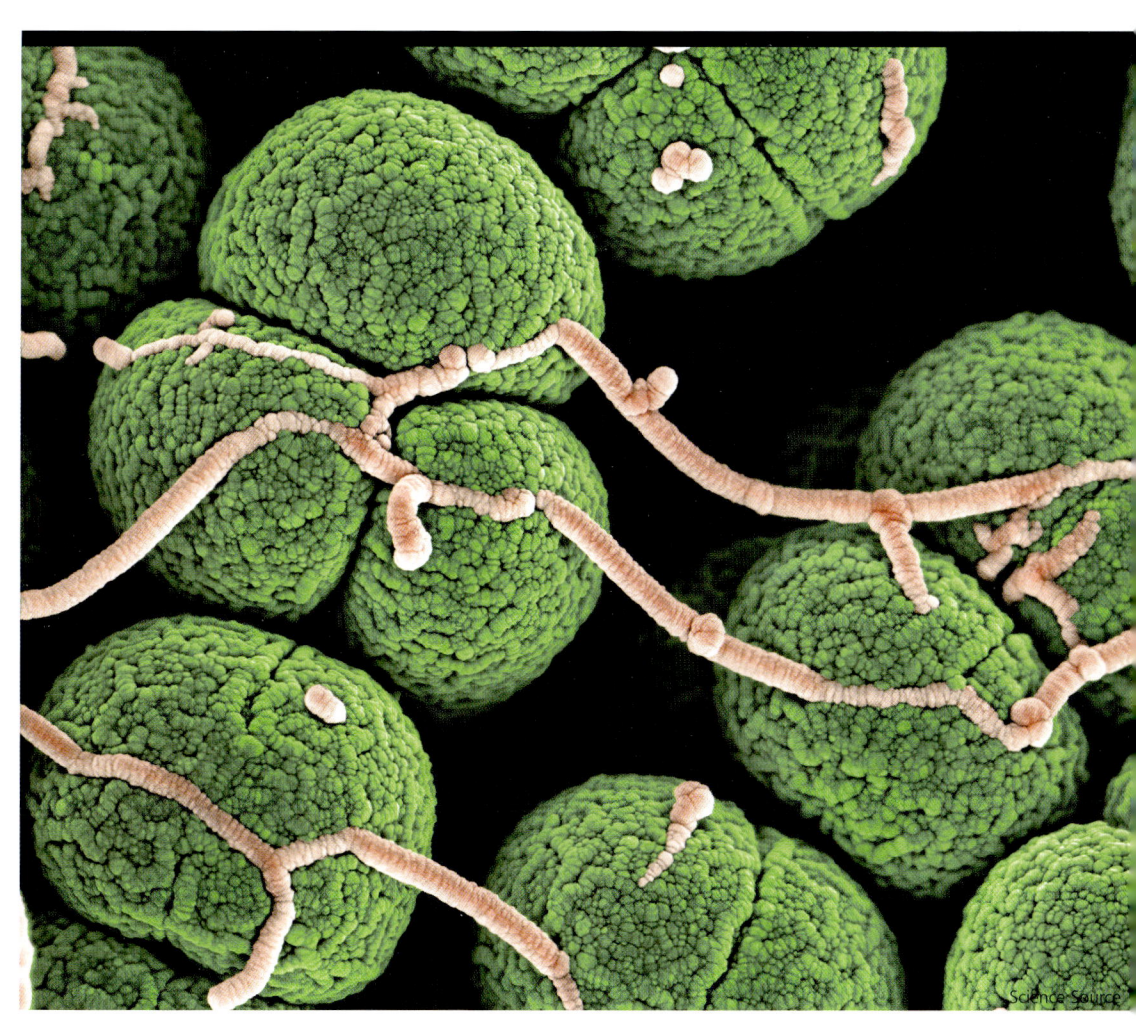

no-till farming 249

Steve Stuebner

Left:
Drew Leitch, left, stands with a seed drilling machine on his farm near Nezperce, Idaho. Kevin Seitz of the Natural Resources Conservation Service shovels a sample of soil.

Below:
Soil microbes may comprise about a fourth of all Earth's species. The weight of those organisms on a foot-deep acre of farmland might exceed 4,800 pounds.

on more than 300,000 acres of farm land in southern Idaho. One of the challenges is that farmers have to dig down 10 to 12 inches to harvest the crop. "You're digging the ground up and moving it around at harvest time," says Wayne Jones, University of Idaho extension agent in Bonneville County. "That makes it problematic."

Scott Steel grows alfalfa, corn, malt barley and other grains on his farm in Bonneville County in eastern Idaho. As he tried to shift to no-till, Steel ran into a common problem in the initial years—weed control following the seeding of a new crop. "One of the biggest things was I was having to use more and more Roundup to control the weeds," Steel says. "I was using way too many chemicals."

Farm experts say this is a common occurrence when shifting to no-till practices. Riggers and Leitch still have to use Roundup to control weeds many years after moving to no-till. Steel also had issues with moving the large direct-seeding farm implement from one field to the next and with fertilizing crops to get the yields he wanted. After six to seven years of trying no-till, he went back to conventional tillage. "My corn and grain crops look better than they did under the no-till system," he says.

In southwest Idaho, no-till is starting to catch on. Local soil and water conservation districts are purchasing no-till drills and renting them to local farmers. There is a long waiting list for the no-till drill owned by the Ada Soil and Water Conservation District. Farm tours of local demonstration projects are helping to spread the word. Brad Brown, who recently retired as an extension crop management specialist for the University of Idaho in Parma, is seeing some direct-seeding activity

South Tyrol Museum of Archaeology

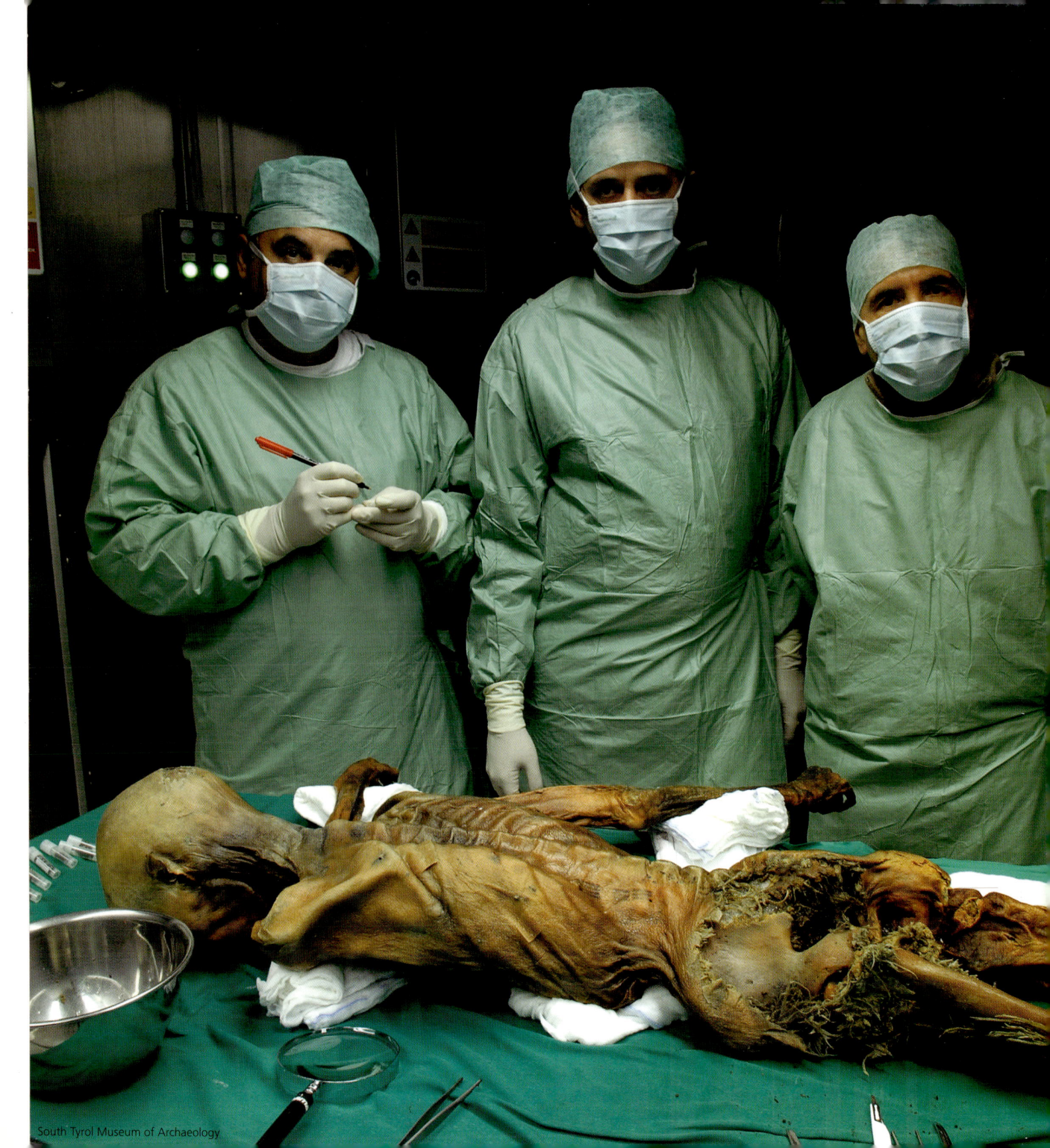

Alien Earth

Microbe hunters have combed only 0.4 percent of the Earth's total mass. Imagine the minutiae of life scientists have yet to explore.

Strata of rock beneath our feet team with organisms. Core samples recovered from coal a mile below the ocean are alive with species that metabolize carbon. The seabed above weaves a briny carpet of plankton through a biosphere that science has studied less closely than the well-mapped surface of Mars. Another terra incognita for science is the stratosphere beyond the rainbow streaked with bacterial dust. In the Italian Alps, meanwhile, ice releases microbes from a mummified human corpse. Ötzi the Iceman—age 5,300, his body pierced by an arrow and found face down in a melting glacier—sustains living type O blood cells and prehistoric fungal spores. The thaw of Ötzi stirs speculation that the planet's ice-bound biota may be a biomass 1,000 times greater than its human population. Species thought extinct may yet come back to life.

Pictured: A mammoth emerging from ice preserves prehistoric microbes; the living floor of the ocean. Left: Italian researchers examine the iceman Ötzi.

Mammoth (Pittsburgh Post-Gazette)

Sea diver (CC BY-SA 2.5)

Acknowledgments

Nature, said William Blake, is heaven in a single flower, a world in a grain of sand. In the making of *Idaho Microbes*, more than 100 contributors saw invisible worlds in grains of Idaho sand. Photographers and artists from six continents created the graphics. Special thanks to Boise colleagues Diane Boothe, Kimberly Catlin, Alicia Dillon, Carl Fritz, John Kelly, Melissa Lavitt, Kathleen Tuck, the College of Education and the Department of History. Pictured: Symmetry helps a bacterium rotate, turning as it swims.

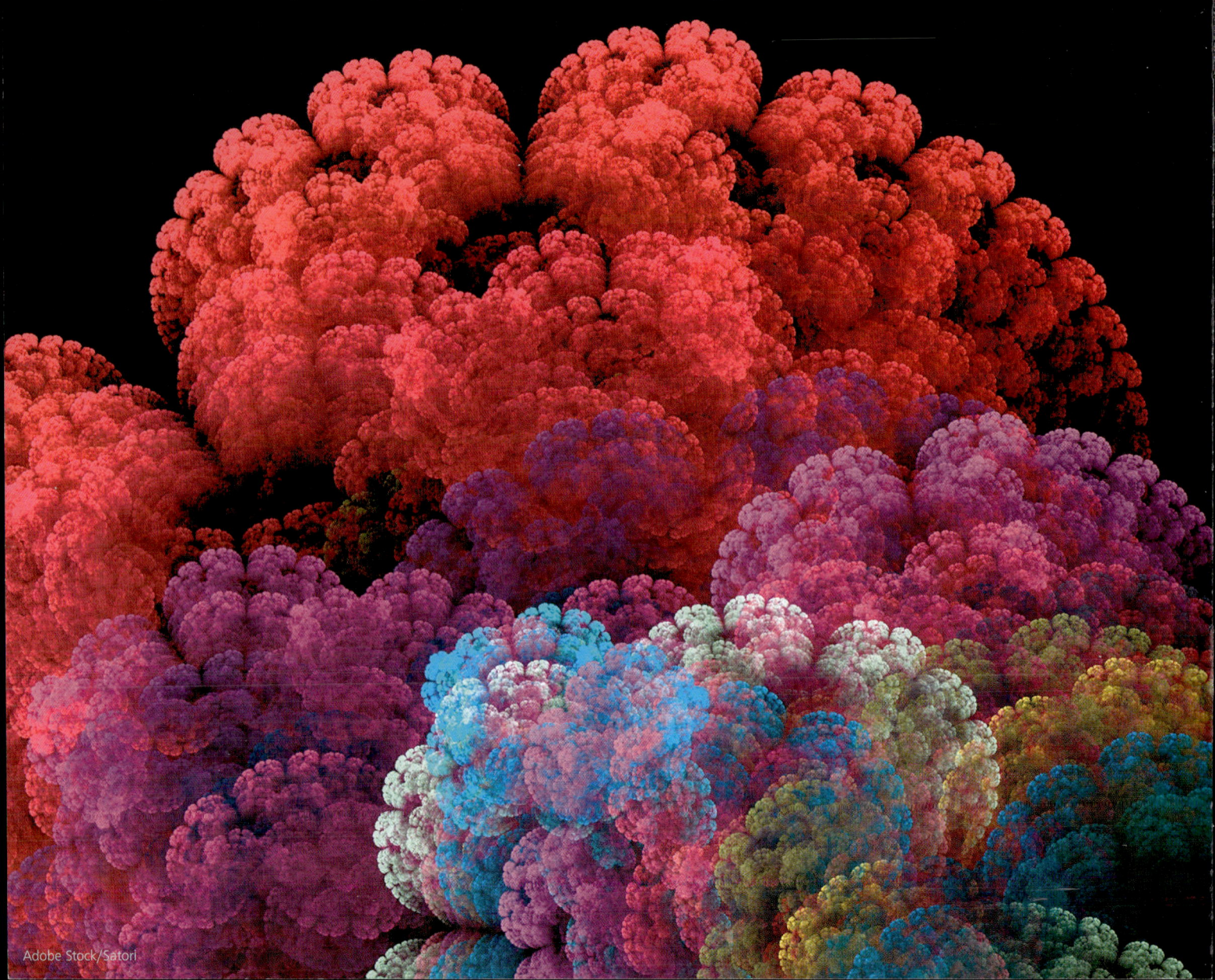

Adobe Stock/Satori

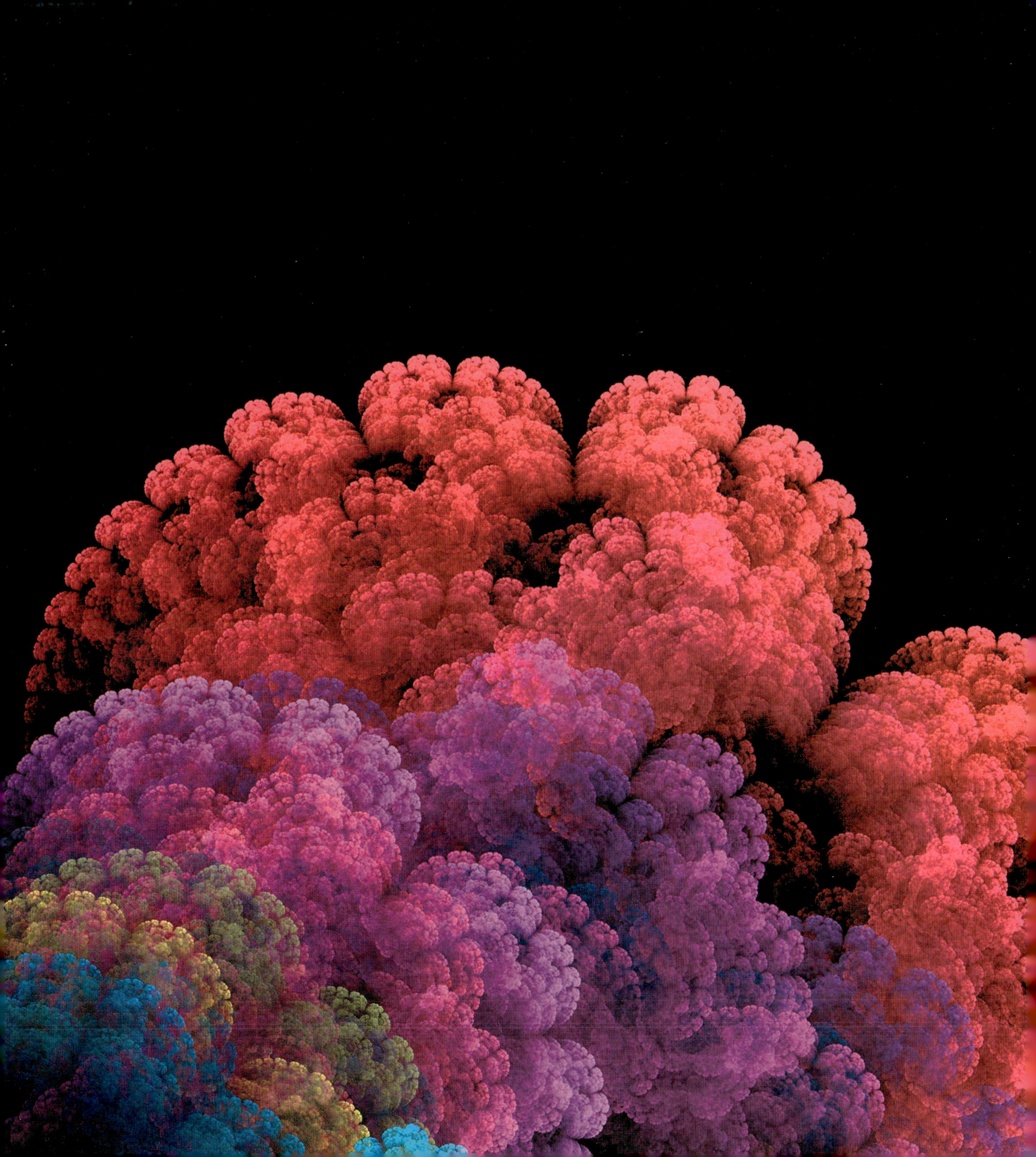

Selected Sources

Amaranthus, Mike, and Bruce Allyn. "Healthy Soil Microbes, Healthy People." *The Atlantic*, June 11, 2013.

Andrews, Michael. *The Life That Lives on Man*. New York: Taplinger, 1976.

Belnap, Jayne, Julie Hilty Kaltenecker, Roger Rosentreter, et al. *Biological Soil Crusts: Ecology and Management*. Technical Reference 1730-2. Washington, DC: U.S. Department of Interior, 2001.

Bench, Molly E., and Merlin M. White. "New Species and First Records of Trichomycetes from Immature Aquatic Insects in Idaho." *Mycologia*, 104: 295-312, 2012.

Biello, David. "Slick Solution: How Microbes Will Clean Up the Deepwater Horizon Oil Spill." *Scientific American*, May 25, 2010.

Blaser, Martin J. *Missing Microbes: How the Overuse of Antibiotics Is Fueling Our Modern Plagues*. New York: Henry Holt, 2014.

Brock, Thomas D. *Thermophilic Microorganisms and Life at High Temperatures*. New York: Springer-Verlag, 1978.

Cedars-Sinai Medical Center. 2014. "Role of Fungus in Digestive Disorders Explored." *Science Daily*, June 6, 2012.

De Kruif, Paul. *The Microbe Hunters*. Originally published in 1926. New York: Houghton Mifflin Harcourt, 2002.

Dusenbery, David B. *Life at Small Scale: The Behavior of Microbes*. New York: Scientific American Library, 1996.

thebottledump.co.uk

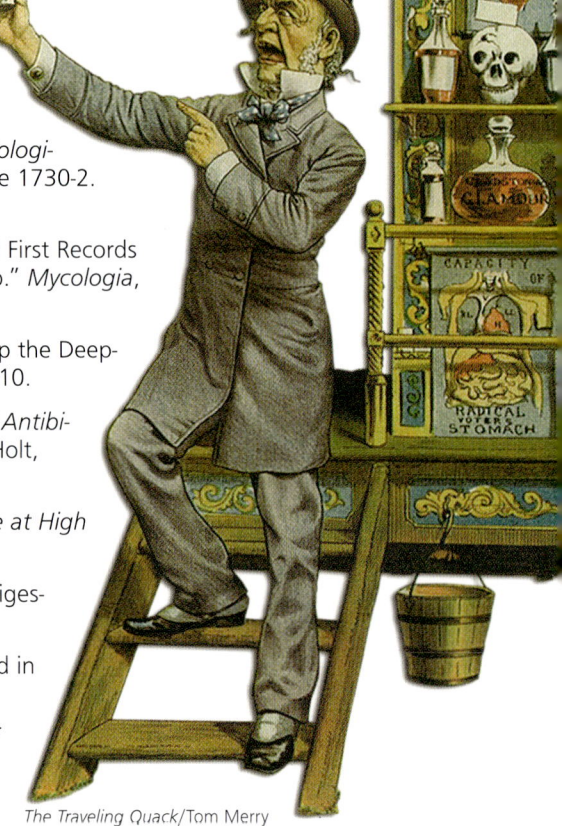

The Traveling Quack/Tom Merry

Left:
The science of Louis Pasteur unleashed a plague of con men and quacks. In 1886, William Radam sold bottles of Microbe Killer that purported to purify blood. Inset: a quack hawks elixirs, 1889.

Below:
An Atlanta company sold brewer's yeast in tablets for weight gain and acne cure. Vitamin B and iron fortified the compound. "Every package of Ionized Yeast is tested and retested biologically," said a 1938 ad.

Environmental Protection Agency. 2001, March. *Cryptosporidium: Drinking Water Health Advisory*. EPA-822-R-01-009. Washington, DC: Office of Science and Technology.

Ferber, Dan. "Minnesota Scientists Develop Bacteria to Clean up Fracking Water." *Midwest Energy News*, September 24, 2012.

Gerardi, Michael H. *The Microbiology of Anaerobic Digesters*. Hoboken, NJ: John Wiley & Sons, 2003.

Ingraham, John L. *March of the Microbes: Sighting the Unseen.* Cambridge, MA: Belknap, 2010.

Keane, Robert E., D. F. Tomback, C. A. Aubry, et al. *A Range-Wide Restoration Strategy for Whitebark Pine (*Pinus albicaulis*)*. General Technical Report RMRS-GTR-279. Boise, ID: Rocky Mountain Research Station, USDA Forest Service, June 2012.

Lantz, G. *Whitebark Pine: An Ecosystem in Peril*. American Forests Special Report, 2010.

Lichtwardt, Robert W., Matías J. Cafaro, and Merlin M. White. *The Trichomycetes: Fungal Associates of Arthropods*, rev. ed. Lawrence: University of Kansas, 2001.

Linn, D. M., and J. W. Doran. "Aerobic and Anaerobic Microbial Populations in No-Till and Plowed Soils." *Soil Science Society of America Journal*, 48: 794-799, 1984.

Luján, Hugo D., and Staffan Svärd, eds. *Giardia: A Model Organism*. New York: Springer, 2011.

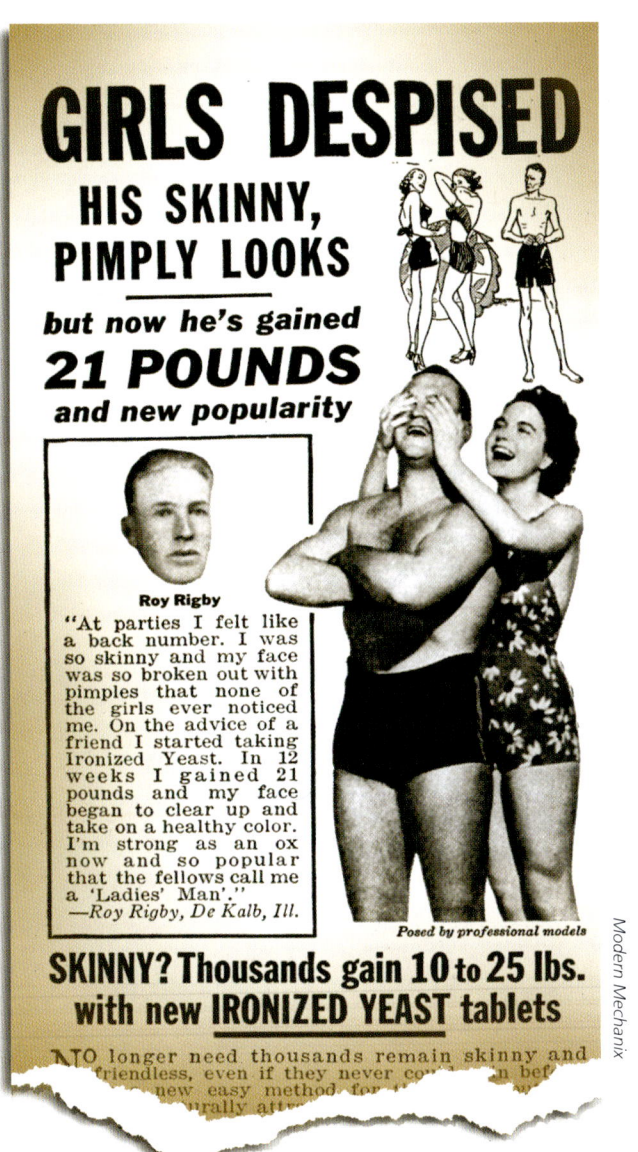

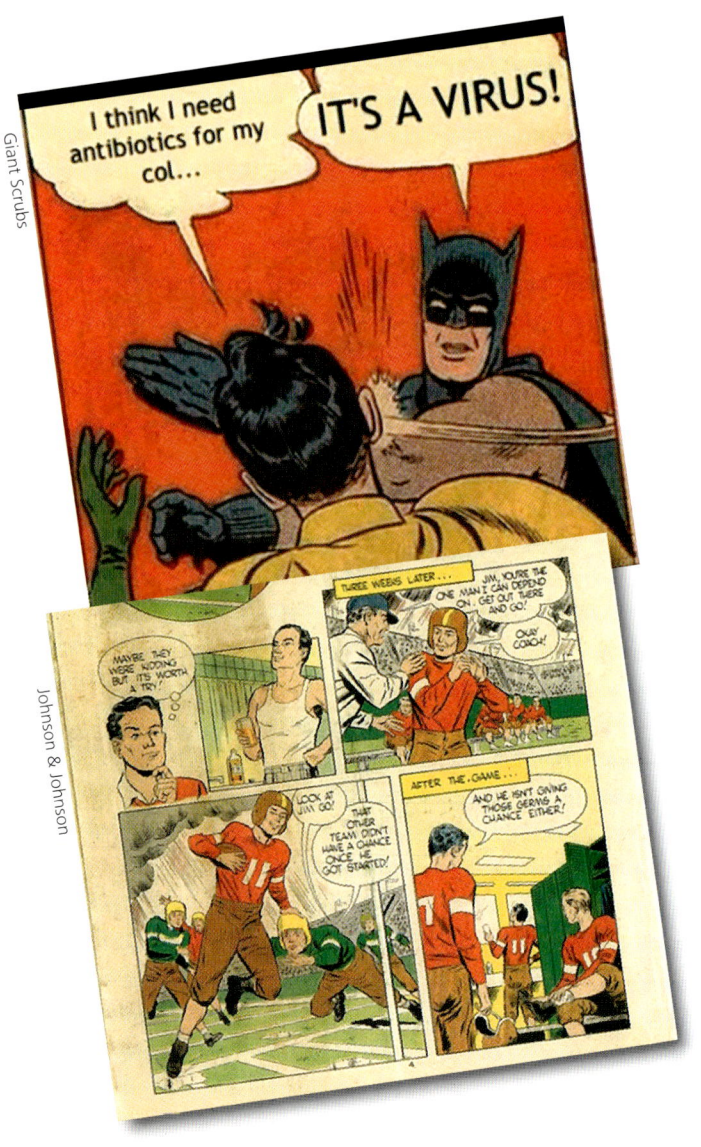

parasites, 125, 129, 135, 195, 247
Parma, 251
Pasteur, Louis, 3, 24, 25, 28, 99, 107, 109, 131, 259
Patterson, David, 30
Payette Brewing, 3, 24, 25, 28, 29, 95, 99, 100, 101, 103, 106, 107, 109, 131, 139, 259
penicillin, 110
Penicillium, 57, 110
Penicillium notatum, 110
Pholiota squarrosa, 63
Pickell, Tyler, 187
Pinus albicaulis (whitebark pine), 145, 155
Poisson, Raymond, 193
polyhydroxyalkanoate, 86
polyploid macronucleus, 28
porcini, 53, 63
prophase, 52
Protista. See protozoa.
protozoa, 1, 48, 87, 193, 206, 231, 247, 248; ciliates, 11, 18–35; parasites, 116, 123, 129, 131
Pseudomonas, 163, 169, 171, 173, 175–177
Pseudomonas fluorescens, 169, 173, 175, 176, 178
Pseudomonas putida, 176, 177
Psora decipiens, 213

Psora montana, 221
Puytoraciella dibryophryis, 27

R
Red Feather (Lounge), 54
Reynolds, Nicole, 187, 204
Rhizopus, 105, 187
Rhizopus stolonifer, 187
Rhopalomyces, 188
Rhopalomyces elegans, 188
Ribes rubrum, 137
Ribes uva-crispa (gooseberry), 153
Riggers, Nathan, 241
Riggers, Steve, 231, 238, 239, 241–244, 246, 251–253
Rock Creek Dairy, 79, 87
Rosentreter, Roger, 211, 213, 214, 216, 217, 221–224, 227, 228
rotifers, 21, 37
Rowan, Eileen, 238

S
Saccharomyces carlsbergensis, 100
Saccharomyces cerevisiae, 93, 97, 105
Saccharomyces pastorianus, 105
Saccharomyces uvarum, 100
sage-grouse, 213, 225–227
Salmon-Challis National Forest, 135

Left:
Batman slaps Robin for a public health announcement. Bottom: A high school hero defeats bacteria in Listerine's Stepping Stones to Success, 1959.

Below:
Sci-fi pioneer Hugo Gernsback developed the germ-free Isolator to keep businessmen hyperfocused, 1925. The oxygen tube blocked airborne pathogens.

Salmon River Brewery, 94, 95, 105, 107
Salmonella, 206
sclerotium, 52
Schwandt, John W., 38, 143
Seitz, Kevin, 238, 245, 251, 252
Selkirk Range, 139
Selway-Bitterroot Wilderness, 139
Smith, Ed, 87
Smith, Phuong, 87
Smiths Ferry, 41
Smittium alpinum, 203
Smittium morbosum, 197, 203
Snake River Birds of Prey National Conservation Area, 23, 216
Snake River Plain, 213, 225
Snow, John, 131
Sockeye Brewing, 95, 105
Spallanzani, Lazzaro, 33
sporangia, 188, 218
Sporosarcina ureae, 249
Stamets, Paul, 48, 53
Stanley, 135, 200
Steele, Mitch, 104
stentors, 36, 37
Sun Valley, 44
Sweet Valley Family Farms, 25, 41, 44, 45, 54, 61, 64
Sweet Valley Organics. See Sweet Valley Family Farms.
Swenson, Paul, 142
symbiosis, 87, 163, 185, 187, 195, 197, 205

T
telia, 135, 145
telophase, 52
Tetrahymena thermophila, 31, 33
Thalassolituus oleivorans, 179
top-fermenting yeasts, 105
Traynor, Eric, 163, 166, 169, 171, 180
Trichomycetes, 187, 190–195, 202, 205
trophozoites, 119, 123, 125, 127
Twin Falls, 72, 77, 87

U
University of Idaho, 58, 86, 148, 213, 225, 243, 251

V
Van Kleek, Mark, 163, 164
Vd'acný, Peter, 33
vermicomposting, 61
Veseth, Roger, 243
Vibrio cholerae, 131
Vorticella, 18

W
Welch, Thomas, 129
Wendell, Laura, 35
West Nile virus, 197
White, Chris, 99
White, Merlin, 25, 187, 203
whitebark pine trees. See *Pinus albicaulis*.
Wicherski, Bruce, 165
Wier, Pete, 149
Wilson, Emma, 195, 202
Winger, Marlon, 235–238, 247–249, 253
Wolfe-Simon, Felisa, 23
Wright, Sara, 247

X
Xanthoria fallax, 225

Y
Yeast, 93, 97–101, 103, 104, 109, 259
Yellowstone National Park, 88, 115, 139, 141, 145
Yoder, Jonathan, 119, 121, 125, 129

Z
Zainasheff, Jamil, 99, 103
Zygomycetes, 187
Zygomycota phylum, 187

About the Author

Steve Stuebner is the award-winning author of more than 10 books on Idaho outdoor topics, including *Cool North Wind* and *Salmon River Country*. He worked for more than 20 years as a journalist, covering stories for the *Idaho Statesman, The New York Times, High Country News, National Wildlife, Columbia Journalism Review* and other publications. Between writing projects, Steve runs a public relations and social media marketing business in Boise, specializing in web video, blogging, web copy, press releases and media relations. He pens a weekly blog, *Stueby's Outdoor Journal*, to share outdoor tips from his hiking, biking and paddling guidebooks.

Stuebner lives in Boise with life partner Wendy Wilson and their blended family of four children. More information at stevestuebner.com